Introducing How Science Works

Science isn't just a long, unchanging list of facts. It's <u>growing</u> and <u>evolving</u> all the time, as scientists develop <u>new theories</u> and <u>gather evidence</u> to test them. Global warming is a good example...

Taking the Temperature of a Planet Is Hard

Years ago, a French scientist worked out that <u>atmospheric gases</u>, including CO_2, keep the Earth at a <u>temperature</u> that's <u>just right</u>. Later, a Swedish chemist, Arrhenius, predicted that as people <u>burnt more coal</u>, the <u>concentration of CO_2</u> in the atmosphere would <u>rise</u>, and the <u>Earth would get warmer</u>.

1) To <u>test</u> this <u>hypothesis</u>, you need <u>reliable data</u> for <u>two variables</u> — the <u>CO_2</u> level and <u>temperature</u>. To be <u>valid</u>, the investigation has to cover the <u>whole globe</u> over <u>hundreds of thousands of years</u> (or we'd just discover that it was colder during the last <u>ice age</u>, which we know anyway).

2) To monitor <u>global temperature</u>, scientists often measure the temperature of the <u>sea surface</u>.

> The first measurements were done from ships — some bloke would fling a <u>bucket</u> overboard, haul it up and stick a <u>thermometer</u> in it. Later, ships recorded the temperature of the water they took on board to cool their engines. Neither method was exactly great —
>
> - Water samples weren't all taken from the same <u>depth</u> (and deeper water is usually <u>colder</u>).
> - The sailors taking the readings were probably a bit <u>slapdash</u> — they were busy <u>sailing</u>.
>   So if two samples were taken in the same place, at the same time, the results would quite possibly be different — in other words, not <u>reproducible</u>.
> - Ships didn't go <u>everywhere</u>, so the records are a bit <u>patchy</u>. So, you might see that the North Atlantic ocean is getting warmer, but have no idea about the rest of the world. The original hypothesis was about <u>global</u> temperature, so the <u>validity</u> of these results is <u>doubtful</u>.

3) Today, things are much better — we can measure sea surface temperature from <u>satellites</u>, with modern, <u>accurate</u> instruments. These results are <u>reliable</u>, and they give us <u>global coverage</u>.

4) We also have very <u>clever</u> ways of finding temperatures and CO_2 levels from the <u>distant past</u> (before thermometers existed) — by examining <u>air bubbles</u> trapped deep in the <u>ice</u> in <u>Antarctica</u>, for example. There are similar tricks involving <u>tree rings</u>, <u>sediments</u> and pollen, so the results can be checked. Even so, these methods <u>aren't perfect</u> — there may be contamination problems, for instance.

Interpreting the Data Is Even Harder...

1) This is a graph of the <u>CO_2</u> and <u>global temperature</u> data. It shows temperature and CO_2 rising very rapidly from about 1850 (when the <u>Industrial Revolution</u> began).

2) <u>But</u> the graph also shows that there have been <u>huge changes</u> in the climate before — you could argue that the recent warming is just part of that <u>natural variability</u>.

3) There's a growing consensus among scientists that the Earth <u>is warming</u>, that it's <u>more</u> than natural variation, and that <u>humans</u> are partly <u>causing</u> it — we're emitting too much CO_2. If that's right, maybe we should <u>stop burning fossil fuels</u>.

4) But there are big <u>interests</u> at stake, and this can influence the way people <u>present</u> the data. If we stopped buying fossil fuels then oil companies, among others, would lose out — so they might emphasise the 'natural variation' argument. People with different interests (like wind turbine manufacturers) might emphasise the more recent rapid rise in temperature and CO_2. They could use exactly the <u>same data</u> — but with a <u>different slant</u>.

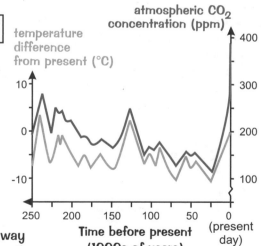

How science works — slowly and painfully...

So, here's how it works — you: 1) observe something (Earth's warming up); 2) come up with a theory to explain it (increasing greenhouse effect); 3) gather <u>valid</u>, <u>reliable evidence</u> to test your theory. If the evidence <u>doesn't</u> match what you predicted, you need to <u>tweak your theory and retest</u>, or <u>start again</u>...

Life and Cells

In physics you often start off with <u>forces</u>, in chemistry it's usually <u>elements</u>, and in biology it's the <u>cell</u>. Not very <u>original</u>, but nice and <u>familiar</u> at least. So away we go — a-one, a-two, a-one, two, three, four...

Plant and Animal Cells have Similarities and Differences

Most <u>human cells</u>, like most <u>animal</u> cells, have the following parts — make sure you know them all:

1) <u>Nucleus</u> — contains <u>genetic material</u> that controls the activities of the cell.

2) <u>Cytoplasm</u> — gel-like substance where most of the <u>chemical reactions</u> happen. It contains <u>enzymes</u> (see page 16) that control these chemical reactions.

3) <u>Cell membrane</u> — holds the cell together and controls what goes <u>in</u> and <u>out</u>.

4) <u>Mitochondria</u> — these are where most of the reactions for <u>respiration</u> take place (see page 17). Respiration releases <u>energy</u> that the cell needs to work.

5) <u>Ribosomes</u> — these are where <u>proteins</u> are made in the cell.

Plant cells usually have <u>all the bits</u> that <u>animal</u> cells have, plus a few <u>extra</u> things that animal cells <u>don't</u> have:

1) Rigid <u>cell wall</u> — made of <u>cellulose</u>. It <u>supports</u> the cell and strengthens it.

2) <u>Permanent vacuole</u> — contains <u>cell sap</u>, a weak solution of sugar and salts.

3) <u>Chloroplasts</u> — these are where <u>photosynthesis</u> occurs, which makes food for the plant (see page 6). They contain a <u>green</u> substance called <u>chlorophyll</u>.

Cells Make Up Tissues, Organs and Systems

Cells have structures that are <u>specialised</u> so they can carry out their <u>function</u> (see next page). Similar cells are grouped together to make a <u>tissue</u>, and different tissues work together as an <u>organ</u>. Organs have a <u>particular job</u> to do in the body — e.g. the <u>heart</u> circulates the blood. Groups of organs working together make up an <u>organ system</u>, like the <u>digestive system</u>. And finally, groups of organs and organ systems working together make up a full <u>organism</u> like you or me. Phew, it's pretty complicated, this life business.

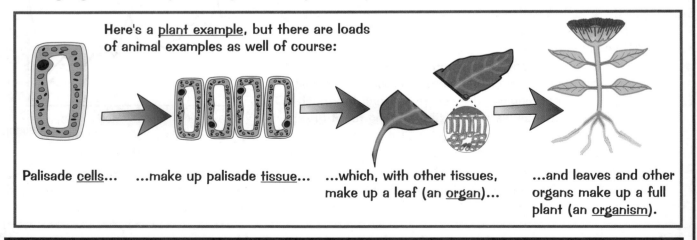

Here's a <u>plant example</u>, but there are loads of animal examples as well of course:

Palisade <u>cells</u>... ...make up palisade <u>tissue</u>... ...which, with other tissues, make up a leaf (an <u>organ</u>)... ...and leaves and other organs make up a full plant (an <u>organism</u>).

There's quite a bit to learn in biology — but that's life, I guess...

At the top of the page are <u>typical cells</u> with all the typical bits you need to know. But cells <u>aren't</u> all the same — they have different <u>structures</u> and <u>produce</u> different substances depending on the <u>job</u> they do.

Specialised Cells

Most cells are specialised for their specific function within a tissue or organ. In the exam you might have to explain how a particular cell is adapted for its function. Here are a few examples that might come up:

1) Palisade Leaf Cells Are Adapted for Photosynthesis

1) Packed with chloroplasts for photosynthesis. More of them are crammed at the top of the cell — so they're nearer the light.
2) Tall shape means a lot of surface area exposed down the side for absorbing CO_2 from the air in the leaf.
3) Thin shape means that you can pack loads of them in at the top of a leaf.

Palisade leaf cells are grouped together to give the palisade layer of a leaf — this is the leaf tissue where most of the photosynthesis happens.

2) Guard Cells Are Adapted to Open and Close Pores

1) Special kidney shape which opens and closes the stomata (pores) in a leaf.
2) When the plant has lots of water the guard cells fill with it and go plump and turgid. This makes the stomata open so gases can be exchanged for photosynthesis.
3) When the plant is short of water, the guard cells lose water and become flaccid, making the stomata close. This helps stop too much water vapour escaping.
4) Thin outer walls and thickened inner walls make the opening and closing work.
5) They're also sensitive to light and close at night to save water without losing out on photosynthesis.

Guard cells are therefore adapted to their function of allowing gas exchange and controlling water loss within the leaf organ.

3) Red Blood Cells Are Adapted to Carry Oxygen

1) Concave shape gives a big surface area for absorbing oxygen. It also helps them pass smoothly through capillaries to reach body cells.
2) They're packed with haemoglobin — the pigment that absorbs the oxygen.
3) They have no nucleus, to leave even more room for haemoglobin.

Red blood cells are an important part of the blood (blood's actually counted as a tissue — weird).

4) Sperm and Egg Cells Are Specialised for Reproduction

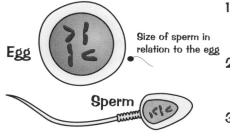

Egg

Size of sperm in relation to the egg

Sperm

1) The main functions of an egg cell are to carry the female DNA and to nourish the developing embryo in the early stages. The egg cell contains huge food reserves to feed the embryo.
2) When a sperm fuses with the egg, the egg's membrane instantly changes its structure to stop any more sperm getting in. This makes sure the offspring end up with the right amount of DNA.
3) The function of a sperm is basically to get the male DNA to the female DNA. It has a long tail and a streamlined head to help it swim to the egg. There are a lot of mitochondria in the cell to provide the energy needed.
4) Sperm also carry enzymes in their heads to digest through the egg cell membrane.

Sperm and eggs are very important cells in the reproductive system.

Beans, flying saucers, tadpoles — cells are masters of disguise...

Okay so the red blood cell doesn't have a nucleus, but apart from that these cells all have all the bits you learnt about on page 2, even though they look completely different and do totally different jobs.

Biology 2(i) — Life Processes

Diffusion

Particles <u>move about randomly</u>, and after a bit they end up <u>evenly spaced</u>. It's not rocket science, is it...

Don't Be Put Off by the Fancy Word

"<u>Diffusion</u>" is simple. It's just the <u>gradual movement</u> of particles from places where there are <u>lots</u> of them to places where there are <u>fewer</u> of them. That's all it is — just the <u>natural tendency</u> for stuff to <u>spread out</u>. Unfortunately you also have to learn the fancy way of saying the same thing, which is this:

> **DIFFUSION is the <u>passive movement</u> of <u>particles</u> from an area of <u>HIGH CONCENTRATION</u> to an area of <u>LOW CONCENTRATION</u>**

Diffusion happens in both <u>liquids</u> and <u>gases</u> — that's because the particles in these substances are free to <u>move about</u> randomly. The <u>simplest type</u> is when different <u>gases</u> diffuse through each other. This is what's happening when the smell of perfume diffuses through a room:

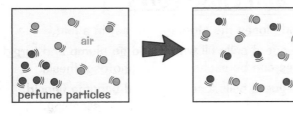

air

perfume particles

smell diffused in the air

The <u>bigger</u> the <u>difference</u> in concentration, the <u>faster</u> the diffusion rate.

Cell Membranes Are Kind of Clever...

They're clever because they <u>hold</u> the cell together <u>BUT</u> they let stuff <u>in and out</u> as well. Substances can move in and out of cells by <u>diffusion</u> and <u>osmosis</u> (see next page). Only very <u>small</u> molecules can <u>diffuse</u> through cell membranes though — things like <u>glucose</u>, <u>amino acids</u>, <u>water</u> and <u>oxygen</u>. <u>Big</u> molecules like <u>starch</u> and <u>proteins</u> can't fit through the membrane.

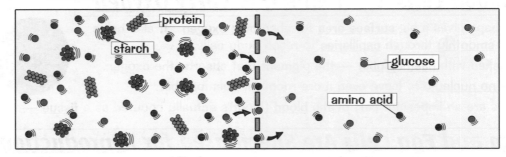

protein

starch

glucose

amino acid

1) Just like with diffusion in air, particles flow through the cell membrane from where there's a <u>high concentration</u> (a lot of them) to where there's a <u>low concentration</u> (not such a lot of them).

2) They're only moving about <u>randomly</u> of course, so they go <u>both</u> ways — but if there are a lot <u>more</u> particles on one side of the membrane, there's a <u>net</u> (overall) movement <u>from</u> that side.

3) The <u>rate</u> of diffusion depends on three main things:

   a) <u>Distance</u> — substances diffuse <u>more quickly</u> when they haven't as <u>far</u> to move. Pretty obvious.

   b) <u>Concentration difference</u> (<u>gradient</u>) — substances diffuse faster if there's a <u>big difference</u> in concentration. If there are <u>lots more</u> particles on one side, there are more there to move across.

   c) <u>Surface area</u> — the <u>more surface</u> there is available for molecules to move across, the <u>faster</u> they can get from one side to the other.

Revision by diffusion — you wish...

Wouldn't that be great — if all the ideas in this book would just gradually drift across into your mind, from an area of <u>high concentration</u> (in the book) to an area of <u>low concentration</u> (in your mind — no offence). Actually, that probably will happen if you read it again. Why don't you give it a go...

Osmosis

If you've got your head round <u>diffusion</u>, osmosis will be a <u>breeze</u>. If not, what are you doing turning over?

Osmosis *is a Special Case* of Diffusion, *That's All*

> <u>OSMOSIS</u> is the <u>movement of water molecules</u> across a <u>partially permeable membrane</u> from a region of <u>high water concentration</u> to a region of <u>low water concentration</u>.

1) A <u>partially permeable</u> membrane is just one with very small holes in it. So small, in fact, only tiny <u>molecules</u> (like water) can pass through them, and bigger molecules (e.g. <u>sucrose</u>) can't.

2) The water molecules actually pass <u>both ways</u> through the membrane during osmosis. This happens because water molecules <u>move about randomly</u> all the time.

3) But because there are <u>more</u> water molecules on one side than on the other, there's a steady <u>net flow</u> of water into the region with <u>fewer</u> water molecules, i.e. into the <u>stronger</u> sugar solution.

4) This means the <u>strong sugar</u> solution gets more <u>dilute</u>. The water acts like it's trying to "<u>even up</u>" the concentration either side of the membrane.

5) Osmosis is a type of <u>diffusion</u> — passive movement of <u>water particles</u> from an area of <u>high water concentration</u> to an area of <u>low water concentration</u>.

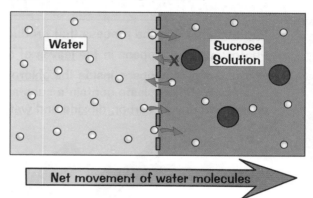

Net movement of water molecules

Water Moves Into and Out of Cells by Osmosis

1) <u>Tissue fluid</u> surrounds the cells in the body — it's basically just <u>water</u> with <u>oxygen</u>, <u>glucose</u> and stuff dissolved in it. It's squeezed out of the <u>blood capillaries</u> to supply the cells with everything they need.

2) The tissue fluid will usually have a <u>different concentration</u> to the fluid <u>inside</u> a cell. This means that water will either move <u>into the cell</u> from the tissue fluid, or <u>out of the cell</u>, by <u>osmosis</u>.

3) If a cell is <u>short of water</u>, the solution inside it will become quite <u>concentrated</u>. This usually means the solution <u>outside</u> is more <u>dilute</u>, and so water will move <u>into</u> the cell by osmosis.

4) If a cell has <u>lots of water</u>, the solution inside it will be <u>more dilute</u>, and water will be <u>drawn out</u> of the cell and into the fluid outside by osmosis.

There's a fairly dull <u>experiment</u> you can do to show osmosis at work.

You cut up an innocent <u>potato</u> into identical cylinders, and get some beakers with <u>different sugar solutions</u> in them. One should be <u>pure water</u>, another should be a <u>very concentrated sugar solution</u>. Then you can have a few others with concentrations <u>in between</u>.

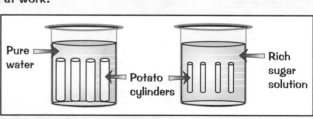

You measure the <u>length</u> of the cylinders, then leave a few cylinders in each beaker for half an hour or so. Then you take them out and measure their lengths <u>again</u>. If the cylinders have drawn in water by osmosis, they'll be a bit <u>longer</u>. If water has been drawn out, they'll have <u>shrunk</u> a bit. Then you can plot a few <u>graphs</u> and things.

The <u>dependent variable</u> is the <u>chip length</u> and the <u>independent variable</u> is the <u>concentration</u> of the sugar solution. All <u>other</u> variables (volume of solution, temperature, time, type of sugar used, etc. etc.) must be kept the <u>same</u> in each case or the experiment won't be a <u>fair test</u>. See, told you it was dull.

And to all you cold-hearted potato murderers...

And that's why it's bad to drink sea-water. The high <u>salt</u> content means you end up with a much <u>lower water concentration</u> in your blood and tissue fluid than in your cells. All the water is sucked out of your cells by osmosis and they <u>shrivel and die</u>. So next time you're stranded at sea, remember this page...

Photosynthesis

You must learn the photosynthesis equation. Learn it so well that you'll still remember it when you're 109.

Learn the Equation for Photosynthesis:

$$\text{Carbon dioxide} + \text{water} \xrightarrow[\text{chlorophyll}]{\text{SUNLIGHT}} \text{glucose} + \text{oxygen}$$

Photosynthesis Produces Glucose Using Sunlight

1) Photosynthesis is the process that produces 'food' in plants. The 'food' it produces is glucose.
2) Photosynthesis happens in the leaves of all green plants — this is largely what the leaves are for.
3) Photosynthesis happens inside the chloroplasts, which are found in leaf cells and in other green parts of a plant. Chloroplasts contain a substance called chlorophyll, which absorbs sunlight and uses its energy to convert carbon dioxide and water into glucose. Oxygen is also produced.

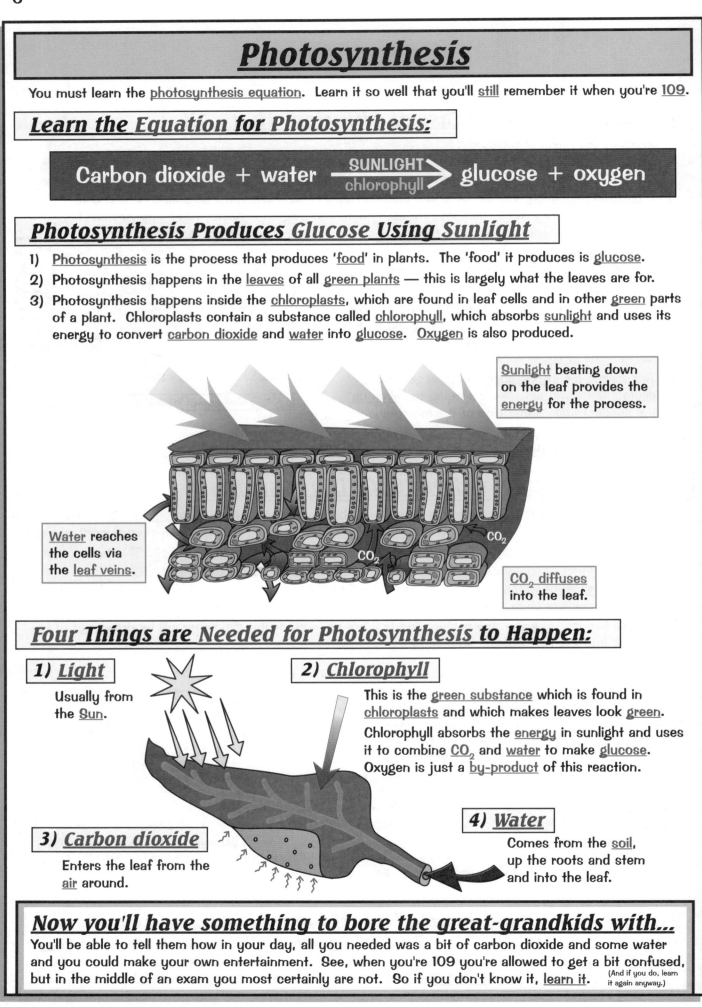

Sunlight beating down on the leaf provides the energy for the process.

Water reaches the cells via the leaf veins.

CO_2

CO_2

CO_2 diffuses into the leaf.

Four Things are Needed for Photosynthesis to Happen:

1) Light
Usually from the Sun.

2) Chlorophyll
This is the green substance which is found in chloroplasts and which makes leaves look green.

Chlorophyll absorbs the energy in sunlight and uses it to combine CO_2 and water to make glucose. Oxygen is just a by-product of this reaction.

3) Carbon dioxide
Enters the leaf from the air around.

4) Water
Comes from the soil, up the roots and stem and into the leaf.

Now you'll have something to bore the great-grandkids with...

You'll be able to tell them how in your day, all you needed was a bit of carbon dioxide and some water and you could make your own entertainment. See, when you're 109 you're allowed to get a bit confused, but in the middle of an exam you most certainly are not. So if you don't know it, learn it. (And if you do, learn it again anyway.)

The Rate of Photosynthesis

The rate of photosynthesis is affected by the amount of <u>light</u>, the amount of <u>CO_2</u>, and the <u>temperature</u>. Plants also need <u>water</u> for photosynthesis, but when a plant is so short of water that it becomes the <u>limiting factor</u> in photosynthesis, it's already in such <u>trouble</u> that this is the least of its worries.

The Limiting Factor Depends on the Conditions

1) Any of these three factors can become the <u>limiting factor</u>. This just means that it's stopping photosynthesis from happening any <u>faster</u>.

2) Which factor is limiting at a particular time depends on the <u>environmental conditions</u>:
   - at <u>night</u> it's pretty obvious that <u>light</u> is the limiting factor,
   - in <u>winter</u> it's often the <u>temperature</u>,
   - if it's warm enough and bright enough, the amount of <u>CO_2</u> is usually limiting.

You can do <u>experiments</u> to work out the <u>ideal conditions</u> for photosynthesis in a particular plant. The easiest type to use is a water plant like <u>Canadian pondweed</u> — you can easily measure the amount of <u>oxygen produced</u> in a given time to show how <u>fast</u> photosynthesis is happening (remember, oxygen is made during photosynthesis).

You could either count the <u>bubbles</u> given off, or if you want to be a bit more <u>accurate</u> you could <u>collect</u> the oxygen in a <u>gas syringe</u>.

— bubbles of oxygen

— pondweed

Three Important Graphs for Rate of Photosynthesis

1) Not Enough Light Slows Down the Rate of Photosynthesis

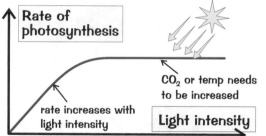

Rate of photosynthesis

CO_2 or temp needs to be increased

rate increases with light intensity

Light intensity

1) Light provides the <u>energy</u> needed for photosynthesis.
2) As the <u>light level</u> is raised, the rate of photosynthesis <u>increases steadily</u> — but only up to a <u>certain point</u>.
3) Beyond that, it <u>won't</u> make any difference because then it'll be either the <u>temperature</u> or the <u>CO_2 level</u> which is the limiting factor.
4) In the lab you can change the light intensity by <u>moving a lamp</u> closer to or further away from your plant.

5) But if you just plot the rate of photosynthesis against "distance of lamp from the beaker", you get a <u>weird-shaped graph</u>. To get a graph like the one above you either need to <u>measure</u> the light intensity at the beaker using a <u>light meter</u> or do a bit of nifty maths with your results.

2) Too Little Carbon Dioxide Also Slows it Down

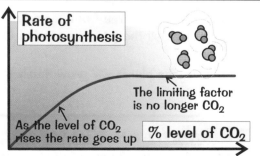

Rate of photosynthesis

The limiting factor is no longer CO_2

As the level of CO_2 rises the rate goes up

% level of CO_2

1) CO_2 is one of the <u>raw materials</u> needed for photosynthesis.
2) As with light intensity the amount of <u>CO_2</u> will only increase the rate of photosynthesis up to a point. After this the graph <u>flattens out</u> showing that CO_2 is no longer the <u>limiting factor</u>.
3) As long as <u>light</u> and <u>CO_2</u> are in plentiful supply then the factor limiting photosynthesis must be <u>temperature</u>.

4) There are loads of different ways to control the amount of CO_2. One way is to dissolve different amounts of <u>sodium hydrogencarbonate</u> in the water, which <u>gives off</u> CO_2.

The Rate of Photosynthesis

3) The Temperature has to be Just Right

Rate of photosynthesis

enzymes destroyed

temperature

45 °C

1) Usually, if the temperature is the limiting factor it's because it's too low — the enzymes needed for photosynthesis work more slowly at low temperatures.

2) But if the plant gets too hot, the enzymes it needs for photosynthesis and its other reactions will be damaged (see p.16).

3) This happens at about 45 °C (which is pretty hot for outdoors, although greenhouses can get that hot if you're not careful).

4) Experimentally, the best way to control the temperature of the flask is to put it in a water bath.

In all these experiments, you have to try and keep all the variables constant apart from the one you're investigating, so it's a fair test:

- use a bench lamp to control the intensity of the light (careful not to block the light with anything)
- keep the flask in a water bath to help keep the temperature constant
- you can't really do anything about the CO_2 levels — you just have to use a large flask, and do the experiments as quickly as you can, so that the plant doesn't use up too much of the CO_2 in the flask. If you're using sodium hydrogencarbonate make sure it's changed each time.

You can Artificially Create the Ideal Conditions for Farming

1) The most common way to artificially create the ideal environment for plants is to grow them in a greenhouse.

2) Greenhouses help to trap the sun's heat, and make sure that the temperature doesn't become limiting. In winter a farmer or gardener might use a heater as well to keep the temperature at the ideal level. In summer it could get too hot, so they might use shades and ventilation to cool things down.

3) Light's always needed for photosynthesis, so commercial farmers often supply artificial light after the Sun goes down to give their plants more quality photosynthesis time.

4) Farmers and gardeners can also increase the level of carbon dioxide in the greenhouse. A fairly common way is to use a paraffin heater to heat the greenhouse. As the paraffin burns, it makes carbon dioxide as a by-product.

5) Keeping plants enclosed in a greenhouse also makes it easier to keep them free from pests and diseases. The farmer can add fertilisers to the soil as well, to provide all the minerals needed for healthy growth (see page 10).

6) If the farmer can keep the conditions just right for photosynthesis, the plants will grow much faster and a decent crop can be harvested much more often.

Don't blame it on the sunshine, don't blame it on the CO_2...

...don't blame it on the temperature, blame it on the plant. Right, and now you'll never forget the three limiting factors in photosynthesis. No... well, make sure you read these pages over and over again till you do. With your newly found knowledge of photosynthesis you could take over the world...

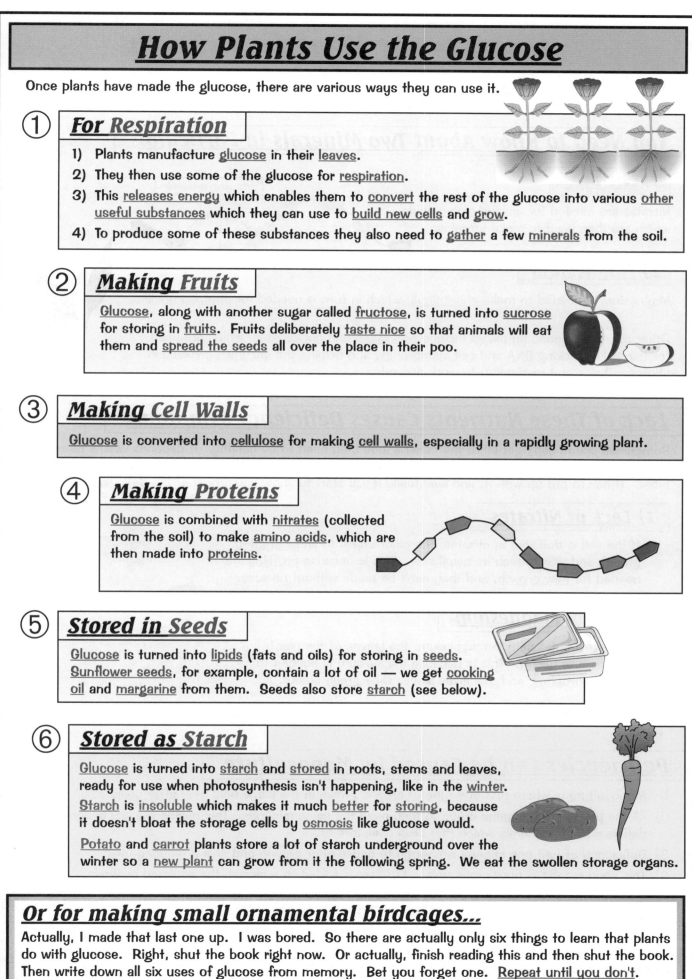

How Plants Use the Glucose

Once plants have made the glucose, there are various ways they can use it.

① For Respiration

1) Plants manufacture glucose in their leaves.
2) They then use some of the glucose for respiration.
3) This releases energy which enables them to convert the rest of the glucose into various other useful substances which they can use to build new cells and grow.
4) To produce some of these substances they also need to gather a few minerals from the soil.

② Making Fruits

Glucose, along with another sugar called fructose, is turned into sucrose for storing in fruits. Fruits deliberately taste nice so that animals will eat them and spread the seeds all over the place in their poo.

③ Making Cell Walls

Glucose is converted into cellulose for making cell walls, especially in a rapidly growing plant.

④ Making Proteins

Glucose is combined with nitrates (collected from the soil) to make amino acids, which are then made into proteins.

⑤ Stored in Seeds

Glucose is turned into lipids (fats and oils) for storing in seeds. Sunflower seeds, for example, contain a lot of oil — we get cooking oil and margarine from them. Seeds also store starch (see below).

⑥ Stored as Starch

Glucose is turned into starch and stored in roots, stems and leaves, ready for use when photosynthesis isn't happening, like in the winter.
Starch is insoluble which makes it much better for storing, because it doesn't bloat the storage cells by osmosis like glucose would.
Potato and carrot plants store a lot of starch underground over the winter so a new plant can grow from it the following spring. We eat the swollen storage organs.

Or for making small ornamental birdcages...

Actually, I made that last one up. I was bored. So there are actually only six things to learn that plants do with glucose. Right, shut the book right now. Or actually, finish reading this and then shut the book. Then write down all six uses of glucose from memory. Bet you forget one. Repeat until you don't.

Minerals for Healthy Growth

Plants need various <u>mineral salts</u>, as well as the carbohydrates they make by photosynthesis, in order to grow properly. They get these mineral ions from the <u>soil</u> by absorbing them through their <u>roots</u>:

You Need to Know About Two Minerals in Particular

1) Nitrates

Nitrates are needed for making <u>amino acids</u>, which are then used to make <u>proteins</u>.

a protein molecule

amino acids

2) Magnesium

Magnesium is needed to make <u>chlorophyll</u>, which in turn is needed for <u>photosynthesis</u>.

Other minerals needed by plants include <u>potassium</u> and <u>phosphates</u>, which are used for things like making <u>DNA</u> and <u>cell membranes</u>, and helping the <u>enzymes</u> involved in photosynthesis and respiration to work properly.

Lack of These Nutrients Causes Deficiency Symptoms

Sometimes plants <u>can't</u> get all of the mineral ions they need to be healthy. It depends what's there in the <u>soil</u> — if the supply of nitrates in the soil gets <u>low</u>, the plant can't just wander off and find some more. It has to put up with it, and eventually it will start to show <u>symptoms</u> of the <u>deficiency</u>.

1) Lack of Nitrates

If the soil is deficient in nitrates, the plant starts to show <u>stunted growth</u> and won't reach its usual size. This is because <u>proteins</u> are needed for <u>new growth</u>, and they can't be made without nitrates.

2) Lack of Magnesium

If the soil is deficient in magnesium, the <u>leaves</u> of the plant start to turn <u>yellow</u>. This is because magnesium is needed to make <u>chlorophyll</u>, and this gives leaves their <u>green</u> colour.

If the plant is left short of the minerals it needs for a long time, it might <u>die</u>.

Deficiencies Can be Caused by Monoculture

1) <u>Monoculture</u> is where just <u>one type of crop</u> is grown in the same field <u>year after year</u>.
2) All the plants are the <u>same</u> crop, so they need the same <u>minerals</u>. This means the soil becomes <u>deficient</u> in the <u>minerals</u> which that crop uses lots of.
3) <u>Deficiency</u> of just <u>one mineral</u> is enough to cause <u>poor growth</u> and give a <u>reduced yield</u>.
4) This soon results in poor crops unless <u>fertiliser</u> is added to <u>replenish</u> the depleted minerals.

Just relax and absorb the information...

When a farmer or a gardener buys <u>fertiliser</u>, that's pretty much what they're buying — a nice big bag of mineral salts to provide all the extra elements plants need to grow. The one they usually need most of is <u>nitrate</u>, which is why manure works quite well — it's full of nitrogenous waste excreted by animals.

Biology 2(i) — Life Processes

Pyramids of Number and Biomass

A <u>trophic level</u> is a <u>feeding</u> level. It comes from the Greek word <u>trophe</u> meaning 'nourishment'. So there.

You Need to Be Able to Construct Pyramids of Biomass

There's <u>less energy</u> and <u>less biomass</u> every time you move <u>up</u> a stage (<u>trophic level</u>) in a food chain. There are usually <u>fewer organisms</u> every time you move up a level too:

<u>100</u> dandelions... feed... <u>10</u> rabbits... which feed... <u>one</u> fox.

This <u>isn't</u> always true though — for example, if <u>500 fleas</u> are feeding on the fox, the number of organisms has <u>increased</u> as you move up that stage in the food chain. So a better way to look at the food chain is often to think about <u>biomass</u> instead of number of organisms. You can use information about biomass to construct a <u>pyramid of biomass</u> to represent the food chain:

1) Each bar on a <u>pyramid of biomass</u> shows the <u>mass of living material</u> at that stage of the food chain — basically how much all the organisms at each level would "<u>weigh</u>" if you put them <u>all together</u>.

2) So the one fox above would have a <u>big biomass</u> and the <u>hundreds of fleas</u> would have a <u>very small biomass</u>. Biomass pyramids are practically <u>always the right shape</u> (unlike number pyramids):

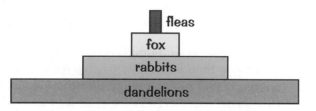

You need to be able to <u>construct</u> pyramids of biomass. Luckily it's pretty simple — they'll give you <u>all</u> the <u>information</u> you need to do it in the exam.

The big bar along the bottom of the pyramid always represents the <u>producer</u> (i.e. a plant). The next bar will be the <u>primary consumer</u> (the animal that eats the plant), then the <u>secondary consumer</u> (the animal that eats the primary consumer) and so on up the food chain. Easy.

You Need to be Able to Interpret Pyramids of Biomass

You also need to be able to look at pyramids of biomass and <u>explain</u> what they show about the <u>food chain</u>. Also very easy. For example:

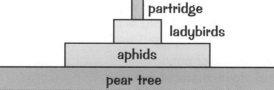

Even if you know nothing about the natural world, you're probably aware that a <u>tree</u> is quite a bit <u>bigger</u> than an <u>aphid</u>. So what's going on here is that <u>lots</u> (probably thousands) of aphids are feeding on a <u>few</u> great big trees. Quite a lot of <u>ladybirds</u> are then eating the aphids, and a few <u>partridges</u> are eating the ladybirds. <u>Biomass</u> and <u>energy</u> are still <u>decreasing</u> as you go up the levels — it's just that <u>one tree</u> can have a very <u>big biomass</u>, and can fix a lot of the <u>Sun's energy</u> using all those leaves.

Constructing pyramids is a breeze — just ask the Egyptians...

There are actually a couple of exceptions where pyramids of <u>biomass</u> aren't quite pyramid-shaped. It happens when the producer has a very short life but reproduces loads, like with plankton at certain times of year. But it's <u>rare</u>, and you <u>don't</u> need to know about it. Forget I ever mentioned it. Sorry.

Energy Transfer and Decay

So now you need to learn <u>why</u> there's <u>less energy</u> and <u>biomass</u> every time you move up a level.

All That Energy Just Disappears Somehow...

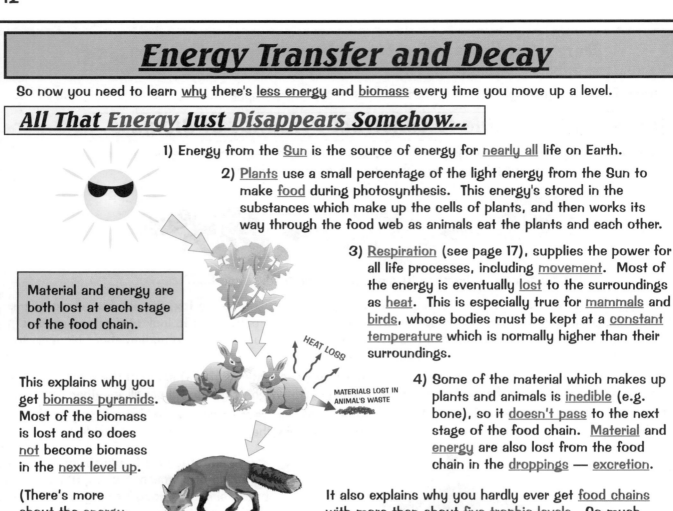

Material and energy are both lost at each stage of the food chain.

HEAT LOSS

MATERIALS LOST IN ANIMAL'S WASTE

1) Energy from the <u>Sun</u> is the source of energy for <u>nearly all</u> life on Earth.

2) <u>Plants</u> use a small percentage of the light energy from the Sun to make <u>food</u> during photosynthesis. This energy's stored in the substances which make up the cells of plants, and then works its way through the food web as animals eat the plants and each other.

3) <u>Respiration</u> (see page 17), supplies the power for all life processes, including <u>movement</u>. Most of the energy is eventually <u>lost</u> to the surroundings as <u>heat</u>. This is especially true for <u>mammals</u> and <u>birds</u>, whose bodies must be kept at a <u>constant temperature</u> which is normally higher than their surroundings.

This explains why you get <u>biomass pyramids</u>. Most of the biomass is lost and so does <u>not</u> become biomass in the <u>next level up</u>.

(There's more about the <u>energy stored</u> in biomass on page 11.)

4) Some of the material which makes up plants and animals is <u>inedible</u> (e.g. bone), so it <u>doesn't pass</u> to the next stage of the food chain. <u>Material</u> and <u>energy</u> are also lost from the food chain in the <u>droppings</u> — <u>excretion</u>.

It also explains why you hardly ever get <u>food chains</u> with more than about <u>five trophic levels</u>. So much energy is <u>lost</u> at each stage that there's not enough left to support more organisms after four or five stages.

Elements are Cycled Back to the Start of the Food Chain by Decay

1) <u>Living things</u> are made of materials they take from the world around them.

2) <u>Plants</u> take elements like <u>carbon</u>, <u>oxygen</u>, <u>hydrogen</u> and <u>nitrogen</u> from the <u>soil</u> or the <u>air</u>. They turn these elements into the <u>complex compounds</u> (carbohydrates, proteins and fats) that make up living organisms, and these then pass through the <u>food chain</u>.

3) These elements are <u>returned</u> to the environment in <u>waste products</u> produced by the organisms, or when the organisms <u>die</u>. These materials decay because they're <u>broken down</u> (digested) by <u>microorganisms</u> — that's how the elements get put back into the <u>soil</u>.

4) Microorganisms work best in <u>warm</u>, <u>moist</u> conditions. Many microorganisms also break down material faster when there's plenty of <u>oxygen</u> available.

5) All the important <u>elements</u> are thus <u>recycled</u> — they return to the soil, ready to be <u>used</u> by new <u>plants</u> and put back into the <u>food chain</u> again.

6) In a <u>stable community</u> the materials <u>taken out</u> of the soil and <u>used</u> are <u>balanced</u> by those that are put <u>back in</u>. There's a constant <u>cycle</u> happening.

Extra decomposers added (compost maker)

Warmth generated by decomposition helps it all along

Compost maker

Finely shredded waste is best

Mesh sides to let air in

So when revising, put the fire on and don't take toilet breaks...

No, I'm being silly, go if you have to. We're talking in <u>general terms</u> about <u>whole food chains</u> here — you won't lose your concentration as a direct result of, erm, excretion.

Managing Food Production

People have been able to use what they know about <u>energy loss</u> from food chains to find the most <u>efficient</u> ways of producing <u>food</u>. But most efficient isn't <u>necessarily</u> best. Although it is often <u>cheapest</u>.

The "Efficiency" of Food Production Can Be Improved...

There are two ways to improve the efficiency of food production:

REDUCE THE NUMBER OF STAGES IN THE FOOD CHAIN

1) For a <u>given area of land</u>, you can produce a lot <u>more food</u> (for humans) by growing <u>crops</u> rather than by having <u>grazing animals</u>. This is because you are <u>reducing</u> the number of <u>stages</u> in the food chain. Only <u>10%</u> of what beef cattle eat becomes <u>useful meat</u> for people to eat.

2) However, people do need to eat a <u>varied diet</u> to stay healthy, and there's still a lot of <u>demand</u> for meat products. Also remember that some land is <u>unsuitable</u> for growing crops, e.g. <u>moorland</u> or <u>fellsides</u>. In these places, animals like <u>sheep</u> and <u>deer</u> might be the <u>best</u> way to get food from the land.

RESTRICT THE ENERGY LOST BY FARM ANIMALS

1) In 'civilised' countries like the UK, animals such as <u>pigs</u> and <u>chickens</u> are often <u>intensively farmed</u>. They're kept <u>close together indoors</u> in small pens, so that they're <u>warm</u> and <u>can't move about</u>.

2) This saves them <u>wasting energy</u> on movement, and stops them giving out so much energy as <u>heat</u>. This makes the <u>transfer of energy</u> from the animal feed to the animal more <u>efficient</u> — so basically, the animals will <u>grow faster</u> on <u>less food</u>.

3) This makes things <u>cheaper</u> for the farmer, and for us when the animals finally turn up on supermarket shelves.

...but it Involves Compromises and Conflict

Improving the efficiency of food production is useful — it means <u>cheaper food</u> for us, and <u>better standards of living</u> for farmers. But it all comes at a <u>cost</u>.

1) Some people think that forcing animals to live in unnatural and uncomfortable conditions is <u>cruel</u>. There's a growing demand for <u>organic meat</u>, which means the animals will <u>not</u> have been intensively farmed.

2) The <u>crowded</u> conditions on factory farms create a favourable environment for the <u>spread of diseases</u>, like avian flu and foot-and-mouth disease.

3) To try to <u>prevent disease</u>, animals are given <u>antibiotics</u>. When the animals are eaten these can enter humans. This allows <u>microbes</u> that infect humans to develop <u>immunity</u> to those antibiotics — so the antibiotics become <u>less effective</u> as <u>human</u> medicines.

4) The environment where the animals are kept needs to be <u>carefully controlled</u>. The animals need to be kept <u>warm</u> to reduce the energy they lose as heat. This often means using power from <u>fossil fuels</u> — which we wouldn't be using if the animals were grazing in their <u>natural</u> environment.

5) Our <u>fish stocks</u> are getting low. Yet a lot of fish goes on feeding animals that are <u>intensively farmed</u> — these animals wouldn't usually eat this source of <u>food</u>.

In an exam, you may be asked to give an account of the <u>positive</u> and <u>negative</u> aspects of food management. You will need to put <u>both sides</u>, whatever your <u>personal opinion</u> is. If you get given some <u>information</u> on a particular case, make sure you <u>use it</u> — they want to see that you've read it <u>carefully</u>.

Locked in a little cage with no sunlight — who'd work in a bank...

You may well have quite a <u>strong opinion</u> on stuff like intensive farming of animals — whether it's 'tree-hugging hippie liberals, just give me a bit of nice cheap pork,' or 'poor creatures, they should be free, free as the wind!' Either way, keep it to yourself and give a nice, <u>balanced argument</u> instead.

The Carbon Cycle

As you've seen, all the nutrients in our environment are constantly being recycled — there's a nice balance between what goes in and what goes out again. This page is all about the recycling of carbon.

The Carbon Cycle Shows How Carbon is Recycled

That can look a bit complicated at first, but it's actually pretty simple.
Learn these important points:

1) There's only one arrow going down from the atmosphere.
The whole thing is "powered" by photosynthesis. CO_2 is removed from the atmosphere by green plants and used to make carbohydrates, fats and proteins in the plants.

2) Some of the CO_2 is returned to the atmosphere when the plants respire.

3) Some of the carbon becomes part of the compounds in animals when the plants are eaten. The carbon then moves through the food chain.

4) Some of the CO_2 is returned to the atmosphere when the animals respire.

5) When plants and animals die, other animals (called detritus feeders) and microorganisms feed on their remains. When these organisms respire, CO_2 is returned to the atmosphere.

6) Animals also produce waste, and this too is broken down by detritus feeders and microorganisms.

7) So the carbon is constantly being cycled — from the air, through food chains and eventually back out into the air again.

Carbon is also released into the atmosphere as CO_2 when plant and animal products are burnt.

What goes around comes around...

And that's the end of the section. But if you were hoping for a nice juicy bit of physics next (as if), or a tasty morsel of chemistry (yeah, right), I'm afraid you're going to be sadly disappointed. Yep, it's more of the same — biology. Oh come on, it could be worse. It could be vomit studies. Or poo analysis.

Revision Summary for Biology 2(i)

And where do you think you're going? It's no use just reading through and thinking you've got it all —
this stuff will only stick in your head if you've learnt it _properly_. And that's what these questions are for.
I won't pretend they'll be easy — they're not meant to be, but all the information's in the section
somewhere. Have a go at all the questions, then if there are any you can't answer, go back, look stuff
up and try again. Enjoy...

1) Name five parts of a cell that both plant and animal cells have. What three things do plant cells have
   that animal cells don't?

2) Name one organ system found in the human body.

3) Give three ways that a palisade leaf cell is adapted for photosynthesis.

4) Give three ways that a sperm cell is adapted for swimming to an egg cell.

5) Name three substances that can diffuse through cell membranes, and two that can't.

6) What three main things does the rate of diffusion depend on?

7) A solution of pure water is separated from a concentrated sugar solution by a partially permeable
   membrane. In which direction will molecules flow, and what substance will these molecules be?

8) An osmosis experiment involves placing pieces of potato into sugar solutions of various
   concentrations and measuring their lengths before and after. What is:
   a) the independent variable,   b) the dependent variable?

9) Write down the equation for photosynthesis.

10) What is the green substance in leaves that absorbs sunlight?

11) Name the three factors that can become limiting in photosynthesis.

12) You carry out an experiment where you change the light intensity experienced by a piece of Canadian
    pondweed by changing the distance between the pondweed and a lamp supplying it with light.
    Write down four things which must be kept constant for this experiment to be a fair test.

13) Explain why it's important that a plant doesn't get too hot.

14) Describe three things that a gardener could do to make sure she gets a good crop of tomatoes.

15) Write down five ways that plants can use the glucose produced by photosynthesis.

16) Why is glucose turned into starch when plants need to store it for later?

17) What is the mineral magnesium needed for in a plant?

18) Describe the symptoms of nitrogen deficiency in a plant.

19) Why are farmers more likely to need extra fertiliser if they grow their crops as a monoculture?

20) What is meant by the term 'biomass'?

21) One oak tree produces acorns that are eaten by ten squirrels. At which stage in this section of the
    food chain is there the greatest:
    a) biomass,   b) energy?

22) Give two ways that energy is lost from a food chain.

23) Explain why mammals and birds tend to lose more energy as heat than reptiles or insects.

24) Why do dead animals and plants decay after they die?

25) A farmer has a field. He plans to grow corn in it and then feed the corn to his cows, which he raises
    for meat. How could the farmer use the field more efficiently to produce food for humans?

26) Why do chickens kept in tiny cages in heated sheds need less food?

27) Summarise the main arguments for and against the intensive farming of animals.

28) Give one way that carbon dioxide from the air enters a food chain.

29) Give three ways that carbon compounds in a food chain become carbon dioxide in the air again.

Biological Catalysts — Enzymes

Chemical reactions are what make you work. And enzymes are what make them work.

Enzymes Are Catalysts Produced by Living Things

1) Living things have thousands of different chemical reactions going on inside them all the time.

2) These reactions need to be carefully controlled — to get the right amounts of substances.

3) You can usually make a reaction happen more quickly by raising the temperature. This would speed up the useful reactions but also the unwanted ones too... not good. There's also a limit to how far you can raise the temperature inside a living creature before its cells start getting damaged.

4) So... living things produce enzymes which act as biological catalysts. Enzymes reduce the need for high temperatures and we only have enzymes to speed up the useful chemical reactions in the body.

> A **CATALYST** is a substance which **INCREASES** the speed of a reaction, without being **CHANGED** or **USED UP** in the reaction.

5) Enzymes are all proteins, which is one reason why proteins are so important to living things.

6) All proteins are made up of chains of amino acids. These chains are folded into unique shapes, which enzymes need to do their jobs (see below).

Enzymes Have Special Shapes So They Can Catalyse Reactions

1) Chemical reactions usually involve things either being split apart or joined together.

2) Every enzyme has a unique shape that fits onto the substance involved in a reaction.

3) Enzymes are really picky — they usually only catalyse one reaction.

4) This is because, for the enzyme to work, the substance has to fit its special shape. If the substance doesn't match the enzyme's shape, then the reaction won't be catalysed.

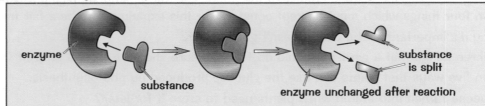

enzyme · substance · substance is split · enzyme unchanged after reaction

Enzymes Need the Right Temperature and pH

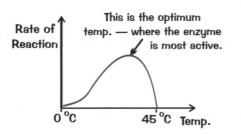

Rate of Reaction — This is the optimum temp. — where the enzyme is most active. — 0 °C — 45 °C Temp.

1) Changing the temperature changes the rate of an enzyme-catalysed reaction.

2) Like with any reaction, a higher temperature increases the rate at first. But if it gets too hot, some of the bonds holding the enzyme together break. This destroys the enzyme's special shape and so it won't work any more. It's said to be denatured.

3) Enzymes in the human body normally work best at around 37 °C — body temperature.

4) The pH also affects enzymes. If it's too high or too low, the pH interferes with the bonds holding the enzyme together. This changes the shape and denatures the enzyme.

Rate of reaction — Optimum pH — pH

5) All enzymes have an optimum pH that they work best at. It's often neutral pH 7, but not always — e.g. pepsin is an enzyme used to break down proteins in the stomach. It works best at pH 2, which means it's well-suited to the acidic conditions there.

If only enzymes could speed up revision...

Just like you've got to have the correct key for a lock, you've got to have the right substance for an enzyme. If the substance doesn't fit, the enzyme won't catalyse the reaction...

Enzymes and Respiration

Many chemical reactions inside cells are controlled by enzymes — including the ones in respiration, protein synthesis and photosynthesis (see page 6).

Enzymes Help Build Amino Acids and Proteins

Enzymes are used to synthesise molecules like amino acids — the ones you don't get from your diet. They also catalyse protein synthesis by joining together amino acids. These proteins could be enzymes — so it all works in a bit of a circle really.

Respiration is NOT "Breathing In and Out"

Respiration involves many reactions, all of which are catalysed by enzymes. These are really important reactions, as respiration releases the energy that the cell needs to do just about everything.

1) Respiration is not breathing in and breathing out, as you might think.

2) Respiration is the process of releasing energy from the breakdown of glucose — and goes on in every cell in your body.

3) It happens in plants too. All living things respire. It's how they release energy from their food.

> **RESPIRATION is the process of RELEASING ENERGY FROM GLUCOSE, which goes on IN EVERY CELL**

Aerobic Respiration Needs Plenty of Oxygen

1) Aerobic respiration is respiration using oxygen. It's the most efficient way to release energy from glucose. (You can also have anaerobic respiration, which happens without oxygen, but that doesn't release nearly as much energy.)

2) Most of the reactions in aerobic respiration happen inside mitochondria (see page 2).

You need to learn the overall word equation for aerobic respiration:

> **Glucose + oxygen ▶ carbon dioxide + water + ENERGY**

Respiration Releases Energy for All Kinds of Things

You need to learn these four examples of what the energy released by aerobic respiration is used for:

1) To build up larger molecules from smaller ones (like proteins from amino acids).

2) In animals, to allow the muscles to contract (which in turn allows them to move about).

3) In mammals and birds the energy is used to keep their body temperature steady (unlike other animals, mammals and birds are warm-blooded).

4) In plants, to build sugars, nitrates and other nutrients into amino acids, which are then built up into proteins.

Breathe, 2, 3, 4 — and release, 6, 7, 8...

So... respiration — that's a pretty important thing. Cyanide is a really nasty toxin that stops respiration by affecting enzymes involved in the process — so it's pretty poisonous (it can kill you). Your brain, heart and liver are affected first because they have the highest energy demands... nice.

Enzymes and Digestion

The enzymes used in <u>respiration</u> work <u>inside cells</u>. Various different enzymes are used in <u>digestion</u> too, but these enzymes are produced by specialised cells and then <u>released</u> into the <u>gut</u> to mix with the food.

Digestive Enzymes Break Down Big Molecules into Smaller Ones

1) <u>Starch</u>, <u>proteins</u> and <u>fats</u> are BIG molecules. They're too big to pass through the walls of the digestive system.

2) <u>Sugars</u>, <u>amino acids</u>, <u>glycerol</u> and <u>fatty acids</u> are much smaller molecules. They can pass easily through the walls of the digestive system.

3) The <u>digestive enzymes</u> break down the BIG molecules into the smaller ones.

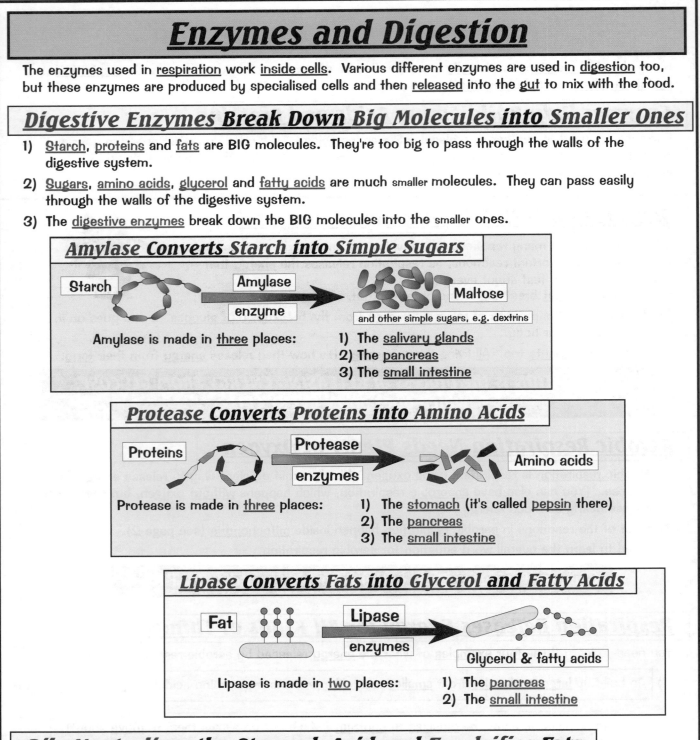

Amylase Converts Starch into Simple Sugars

Starch → Amylase enzyme → Maltose
and other simple sugars, e.g. dextrins

Amylase is made in <u>three</u> places:
1) The <u>salivary glands</u>
2) The <u>pancreas</u>
3) The <u>small intestine</u>

Protease Converts Proteins into Amino Acids

Proteins → Protease enzymes → Amino acids

Protease is made in <u>three</u> places:
1) The <u>stomach</u> (it's called <u>pepsin</u> there)
2) The <u>pancreas</u>
3) The <u>small intestine</u>

Lipase Converts Fats into Glycerol and Fatty Acids

Fat → Lipase enzymes → Glycerol & fatty acids

Lipase is made in <u>two</u> places:
1) The <u>pancreas</u>
2) The <u>small intestine</u>

Bile Neutralises the Stomach Acid and Emulsifies Fats

1) Bile is <u>produced</u> in the <u>liver</u>. It's <u>stored</u> in the <u>gall bladder</u> before it's released into the <u>small intestine</u>.

2) The <u>hydrochloric acid</u> in the stomach makes the pH <u>too acidic</u> for enzymes in the small intestine to work properly. Bile is <u>alkaline</u> — it <u>neutralises</u> the acid and makes conditions <u>alkaline</u>. The enzymes in the small intestine <u>work best</u> in these alkaline conditions.

3) It <u>emulsifies</u> fats. In other words it breaks the fat into <u>tiny droplets</u>. This gives a much <u>bigger surface area</u> of fat for the enzyme lipase to work on — which makes its digestion <u>faster</u>.

What do you call an acid that's eaten all the pies...

This all happens inside our digestive system, but there are some microorganisms which secrete their digestive enzymes <u>outside their body</u> onto the food. The food's digested, then the microorganism absorbs the nutrients. Nice. I wouldn't like to empty the contents of my stomach onto my plate before eating it.

The Digestive System

So now you know what the enzymes do, here's a nice big picture of the whole of the digestive system.

The Breakdown of Food is Catalysed by Enzymes

1) Enzymes used in the digestive system are produced by specialised cells in glands and in the gut lining.
2) Different enzymes catalyse the breakdown of different food molecules.

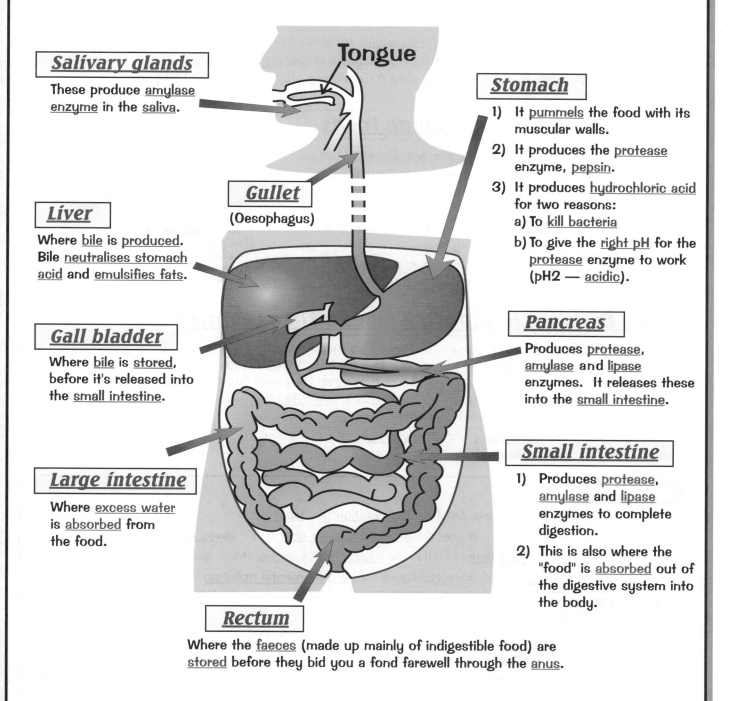

Tongue

Salivary glands
These produce amylase enzyme in the saliva.

Gullet
(Oesophagus)

Liver
Where bile is produced. Bile neutralises stomach acid and emulsifies fats.

Gall bladder
Where bile is stored, before it's released into the small intestine.

Large intestine
Where excess water is absorbed from the food.

Rectum
Where the faeces (made up mainly of indigestible food) are stored before they bid you a fond farewell through the anus.

Stomach
1) It pummels the food with its muscular walls.
2) It produces the protease enzyme, pepsin.
3) It produces hydrochloric acid for two reasons:
   a) To kill bacteria
   b) To give the right pH for the protease enzyme to work (pH2 — acidic).

Pancreas
Produces protease, amylase and lipase enzymes. It releases these into the small intestine.

Small intestine
1) Produces protease, amylase and lipase enzymes to complete digestion.
2) This is also where the "food" is absorbed out of the digestive system into the body.

Mmmm — so who's for a chocolate digestive...

Did you know that the whole of your digestive system is actually a hole that goes right through your body. Think about it. It just gets loads of food, digestive juices and enzymes piled into it. Most of it's then absorbed back into the body and the rest is politely stored ready for removal.

Uses of Enzymes

Some microorganisms produce enzymes which pass <u>out</u> of their cells and catalyse reactions outside them (e.g. to <u>digest</u> the microorganism's <u>food</u>). These enzymes have many <u>uses</u> in the <u>home</u> and in <u>industry</u>.

Enzymes Are Used in *Biological Detergents*

1) <u>Enzymes</u> are the '<u>biological</u>' ingredients in biological detergents and washing powders.
2) They're mainly <u>protein-digesting</u> enzymes (proteases) and <u>fat-digesting</u> enzymes (lipases).
3) Because the enzymes attack <u>animal</u> and <u>plant</u> matter, they're ideal for removing <u>stains</u> like <u>food</u> or <u>blood</u>.

Enzymes Are Used to *Change Foods*

1) The <u>proteins</u> in some <u>baby foods</u> are '<u>pre-digested</u>' using protein-digesting enzymes (<u>proteases</u>), so they're easier for the baby to digest.
2) Carbohydrate-digesting enzymes (<u>carbohydrases</u>) can be used to turn <u>starch syrup</u> (yuk) into <u>sugar syrup</u> (yum).
3) <u>Glucose syrup</u> can be turned into <u>fructose syrup</u> using an isomerase enzyme. Fructose is <u>sweeter</u>, so you can use <u>less</u> of it — good for slimming foods and drinks.

Using Enzymes in *Industry Takes a Lot of Control*

Enzymes are <u>really useful</u> in industry. They <u>speed up</u> reactions without the need for <u>high temperatures</u> and <u>pressures</u>. In a big industrial plant the substances are often continually run over the enzymes, so they have to be kept from <u>washing away</u>. They can be trapped in an <u>alginate bead</u> (a bead of jelly-like stuff) or in a latticework of <u>silica gel</u>.

You need to know the <u>advantages</u> and <u>disadvantages</u> of using them, so here are a few to get you started:

ADVANTAGES

1) They're <u>specific</u>, so they only catalyse the <u>reaction</u> you <u>want</u> them to.
2) Using lower temperatures and pressures means a <u>lower cost</u> and it <u>saves energy</u>.
3) Enzymes work for a <u>long time</u>, so after the <u>initial cost</u> of buying them, you can <u>continually</u> use them.
4) They are <u>biodegradable</u> and therefore cause less <u>environmental pollution</u>.

DISADVANTAGES

1) Some people can develop <u>allergies</u> to the enzymes (e.g. in biological washing powders).
2) Enzymes can be <u>denatured</u> by even a <u>small</u> increase in temperature. They're also susceptible to <u>poisons</u> and changes in <u>pH</u>. This means the conditions in which they work must be <u>tightly controlled</u>.
3) <u>Contamination</u> of the enzyme with other substances can affect the reaction.

There's a lot to learn — but don't be deterred gents...

Enzymes are so <u>picky</u>. Even tiny little changes in pH or temperature will stop them working at maximum efficiency. They only catalyse <u>one reaction</u> as well, so you need to use a different one for each reaction, not like using transition metals as catalysts (see page 59). Temperamental little things these enzymes...

Homeostasis

Homeostasis is a fancy word. It covers lots of things, so I guess it has to be. Homeostasis covers all the functions of your body which try to maintain a "constant internal environment". Learn that definition:

HOMEOSTASIS is the maintenance of a constant internal environment.

There Are Six Main Things That Need to Be Controlled

The first four are all things you need, but at just the right level — not too much and not too little.

1) The body temperature can't get too hot or too cold (see below).
2) Water content mustn't get too high or low, or too much water could move into or out of cells and damage them. There's more on controlling water content on page 22.
3) If the ion content of the body is wrong, the same thing could happen. See page 22.
4) The blood sugar level needs to stay within certain limits (see page 23).

The last two are waste products — they're constantly produced in the body and you need to get rid of them.

5) Carbon dioxide is a product of respiration. It's toxic in high quantities so it's got to be removed. It leaves the body by the lungs when you breathe out.
6) Urea is a waste product made from excess amino acids. There's more about it below.

Body Temperature Must Be Carefully Controlled

All enzymes work best at a certain temperature (see page 16). The enzymes within the human body work best at about 37 °C. If the body gets too hot or too cold, the enzymes won't work properly and some really important reactions could be disrupted. In extreme cases, this can even lead to death.

1) There is a thermoregulatory centre in the brain which acts as your own personal thermostat.
2) It contains receptors that are sensitive to the temperature of the blood flowing through the brain.
3) The thermoregulatory centre also receives impulses from the skin, giving info about skin temperature.
4) If you're getting too hot or too cold, your body can respond to try and cool you down or warm you up:

When you're TOO HOT:

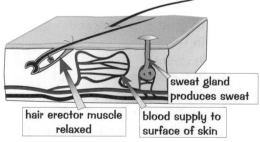

sweat gland produces sweat

hair erector muscle relaxed | blood supply to surface of skin

1) Hairs lie flat.
2) Sweat is produced by sweat glands and evaporates from the skin, which removes heat.
3) The blood vessels supplying the skin dilate so more blood flows close to the surface of the skin. This makes it easier for heat to be transferred from the blood to the environment.

When you're TOO COLD:

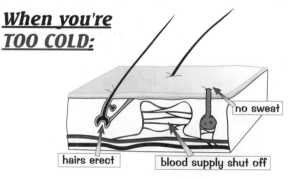

no sweat

hairs erect | blood supply shut off

1) Hairs stand up to trap an insulating layer of air.
2) No sweat is produced.
3) Blood vessels supplying skin capillaries constrict to close off the skin's blood supply.

When you're cold you shiver too (your muscles contract automatically). This needs respiration, which releases some energy as heat.

Shiver me timbers — it's a wee bit nippy in here...

People who are exposed to extreme cold for long periods of time without protection can get frostbite — the blood supply to the fingers and toes is cut off to conserve heat (but this kills the cells, and they go black)...yuk

The Kidneys and Homeostasis

Kidneys are really important in this whole homeostasis thing.

Kidneys Basically Act as Filters to "Clean the Blood"

The <u>kidneys</u> perform <u>three main roles</u>:

1) <u>Removal of urea</u> from the blood.
2) <u>Adjustment of ions</u> in the blood.
3) <u>Adjustment of water content</u> of the blood.

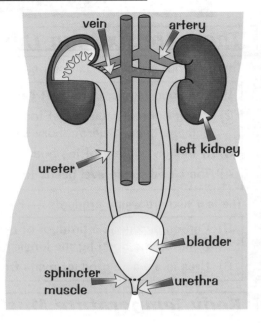

1) Removal of Urea

1) Proteins can't be <u>stored</u> by the body — so any excess amino acids are converted into <u>fats</u> and <u>carbohydrates</u>, which can be stored.

2) This process occurs in the <u>liver</u>. <u>Urea</u> is produced as a <u>waste product</u> from the reactions.

3) Urea is <u>poisonous</u>. It's released into the <u>bloodstream</u> by the liver. The <u>kidneys</u> then filter it out of the blood and it's excrete from the body in <u>urine</u>.

2) Adjustment of Ion Content

1) <u>Ions</u> such as <u>sodium</u> are taken into the body in <u>food</u>, and then absorbed into the blood.

2) If the ion content of the body is <u>wrong</u>, this could mean too much or too little <u>water</u> is drawn into cells by <u>osmosis</u> (see page 5). Having the wrong amount of water can <u>damage</u> cells.

3) Excess ions are <u>removed</u> by the kidneys. For example, a salty meal will contain far too much sodium and so the kidneys will remove the <u>excess</u> sodium ions from the blood.

4) Some ions are also lost in <u>sweat</u> (which tastes salty, you may have noticed).

5) But the important thing to remember is that the <u>balance</u> is always maintained by the <u>kidneys</u>.

3) Adjustment of Water Content

Water is taken into the body as <u>food and drink</u> and is <u>lost</u> from the body in <u>three main ways</u>: 1) In <u>urine</u>
    2) In <u>sweat</u>
    3) In the air we <u>breathe out</u>.

The body has to <u>constantly balance</u> the water coming in against the water going out. Our bodies can't control how much we lose in our breath, but we do control the other factors. This means the <u>water balance</u> is between: 1) Liquids <u>consumed</u>
    2) Amount <u>sweated out</u>
    3) Amount <u>excreted by the kidneys</u> in the <u>urine</u>.

On a <u>cold</u> day, if you <u>don't sweat</u>, you'll produce <u>more urine</u> which will be <u>pale</u> and <u>dilute</u>.

On a <u>hot</u> day, you <u>sweat a lot</u>, and you'll produce <u>less urine</u> which will be <u>dark-coloured</u> and <u>concentrated</u>.

The water lost when it is hot has to be <u>replaced</u> with water from food and drink to restore the <u>balance</u>.

Adjusting water content — blood, sweat and, erm, wee...

Scientists have made a machine which can do the kidney's job for us — a <u>kidney dialysis machine</u>. People with kidney failure have to use it for 3-4 hours, 3 times a week. Unfortunately it's not something you can carry around in your back pocket, which makes life difficult for people with kidney failure.

Controlling Blood Sugar

Blood sugar is also controlled as part of homeostasis. Insulin is a hormone that controls how much sugar there is in your blood. Learn how it does it:

Insulin Controls Blood Sugar Levels

1) Eating foods containing carbohydrate puts glucose into the blood from the gut.

2) Normal metabolism (reactions) of cells removes glucose from the blood.

3) Vigorous exercise also removes a lot of glucose from the blood.

4) Levels of glucose in the blood must be kept steady. Changes in blood glucose are monitored and controlled by the pancreas, using the hormone insulin, as shown:

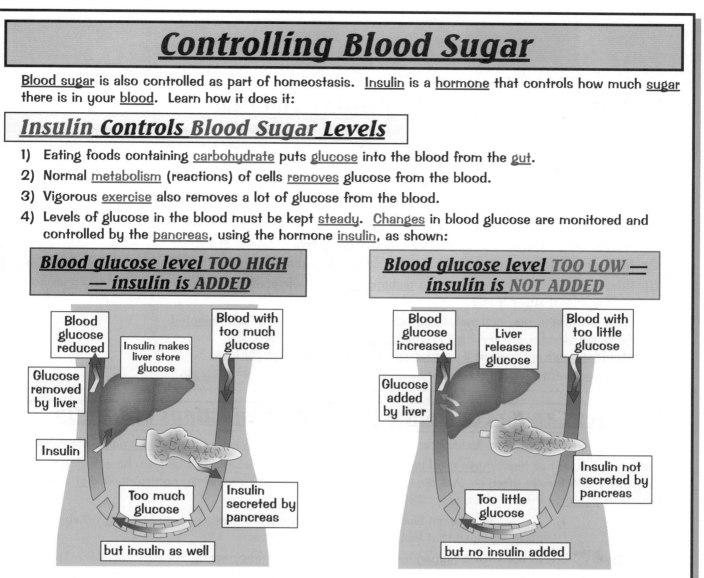

Blood glucose level TOO HIGH — insulin is ADDED

Blood glucose reduced

Insulin makes liver store glucose

Blood with too much glucose

Glucose removed by liver

Insulin

Too much glucose

Insulin secreted by pancreas

but insulin as well

Blood glucose level TOO LOW — insulin is NOT ADDED

Blood glucose increased

Liver releases glucose

Blood with too little glucose

Glucose added by liver

Too little glucose

Insulin not secreted by pancreas

but no insulin added

Diabetes (type 1) — the Pancreas Stops Making Enough Insulin

1) Diabetes (type 1) is a disorder where the pancreas doesn't produce enough insulin.

2) The result is that a person's blood sugar can rise to a level that can kill them.

3) The problem can be controlled in two ways:

   a) Avoiding foods rich in simple carbohydrates, e.g. sugars (which cause glucose levels to rise rapidly). It can also be helpful to take exercise after eating carbohydrates... i.e. trying to use up the extra glucose by doing physical activity — but this isn't usually very practical.

   b) Injecting insulin into the blood at mealtimes. This will make the liver remove the glucose as soon as it enters the blood from the gut, when the food is being digested. This stops the level of glucose in the blood from getting too high and is a very effective treatment. However, the person must make sure they eat sensibly after injecting insulin, or their blood sugar could drop dangerously.

4) The amount of insulin that needs to be injected depends on the person's diet and how active they are.

5) Diabetics can check their blood sugar using a glucose-monitoring device. This is a little hand-held machine. They prick their finger to get a drop of blood for the machine to check.

My blood sugar feels low after all that — pass the biscuits...

This stuff can seem a bit confusing at first, but if you concentrate on learning those two diagrams, it'll all start to get a lot easier. Don't forget that only carbohydrate foods put the blood sugar levels up.

Insulin and Diabetes

Scientific discoveries often take a long time, and a lot of trial and error — here's a rather famous example.

Insulin Was Discovered by Banting and Best

It has been known for some time that people who suffer from diabetes have a lot of <u>sugar</u> in their <u>urine</u>. In the 19th century, scientists <u>removed pancreases</u> from dogs, and the same sugary urine was observed — the dogs became <u>diabetic</u>. That suggested that the pancreas had to have something to do with the illness. In the 1920s Frederick <u>Banting</u> and his assistant Charles <u>Best</u> managed to successfully <u>isolate insulin</u> — the hormone that controls blood sugar levels.

1) Banting and Best <u>tied string</u> around a dog's pancreas so that a lot of the organ <u>wasted</u> away — but the bits which made the <u>hormones</u> were left <u>intact</u>.
2) They <u>removed</u> the pancreas from the dog, and obtained an <u>extract</u> from it.
3) They then injected this extract into <u>diabetic dogs</u> and observed the effects on their <u>blood sugar levels</u>.
4) From the graph, you can see that after the pancreatic extract was <u>injected</u>, the dog's blood sugar level <u>fell dramatically</u>. This showed that the <u>pancreatic extract</u> caused a <u>temporary decrease</u> in <u>blood sugar level</u>.
5) They went on to <u>isolate</u> the substance in the pancreatic extract — <u>insulin</u>.

Diabetes Can Be Controlled by Regular Injections of Insulin

After <u>a lot</u> more experiments, Banting and Best tried <u>injecting insulin</u> into a <u>diabetic human</u>. And it <u>worked</u>. Since then insulin has been <u>mass produced</u> to meet the <u>needs</u> of diabetics. Diabetics have to inject themselves with insulin <u>often</u> — 2-4 times a day. They also need to carefully control their <u>diet</u> and the amount of <u>exercise</u> they do (see page 23).

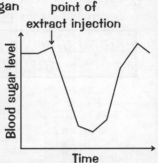

1) At first, the insulin was extracted from the pancreases of <u>pigs</u> or <u>cows</u>. Diabetics used <u>glass syringes</u> that had to be boiled before use.
2) In the 1980s <u>human</u> insulin made by <u>genetic engineering</u> became available. This didn't cause any <u>adverse reactions</u> in patients, which <u>animal</u> insulin sometimes did.
3) <u>Slow</u>, <u>intermediate</u> and <u>fast</u> acting insulins have been developed to make it easier for diabetics to <u>control</u> their blood sugar levels.
4) Ready sterilised, <u>disposable syringes</u> are now available, as well as <u>needle-free devices</u>.

Improving methods of treatment allow diabetics to <u>control</u> their blood sugar <u>more easily</u>. This helps them avoid some of the damaging side effects of poor control, such as <u>blindness</u> and <u>gangrene</u>.

Diabetics May Have a Pancreas Transplant

Injecting yourself with insulin every day <u>controls</u> the effects of diabetes, but it doesn't help to cure it.

1) Diabetics can have a <u>pancreas transplant</u>. A successful operation means they won't have to inject themselves with insulin again. But as with any organ transplant, your body can <u>reject</u> the tissue. If this happens you have to take <u>costly immunosuppressive drugs</u>, which often have <u>serious side-effects</u>.
2) Another method, still in its <u>experimental stage</u>, is to transplant just the <u>cells</u> which produce insulin. There's been <u>varying success</u> with this technique, and there are still problems with <u>rejection</u>.
3) Modern research into <u>artificial pancreases</u> and <u>stem cell research</u> may mean the elimination of organ rejection, but there's a way to go yet (see page 29).

Blimey — all that in the last hundred years...

Insulin can't be taken in a pill or tablet — the <u>enzymes</u> in the stomach completely <u>destroy it</u> before it reaches the bloodstream. That's why diabetics have to <u>inject it</u>. Diabetes is becoming more and more common, partly due to our society becoming increasingly overweight. It's very serious.

Revision Summary for Biology 2(ii)

There are two quite separate bits to this section. First you've got enzymes, how they're used inside cells, in digestion, and in industry. Then the second bit is all about keeping things constant in your body. Have a bash at the questions, go back and check anything you're not sure about, then try again. Practise until you can answer all these questions really easily without having to look back at the section.

1) Give a definition of a catalyst.

2) Name three enzyme-catalysed chemical reactions that happen inside living organisms.

3) What family of molecules do enzymes belong to?

4) Explain why an enzyme-catalysed reaction stops when the reaction mixture is heated above a certain temperature.

5)* The graph on the right shows how the rate of an enzyme-catalysed reaction depends on pH:
   a) State the optimum pH of the enzyme.
   b) In which part of the human digestive system would you expect to find the enzyme?

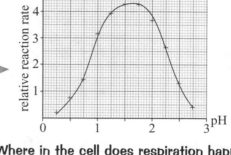

6) In which cells of the body does respiration happen? Where in the cell does respiration happen?

7) Write down the word equation for aerobic respiration.

8) Give three examples of life processes for which the energy from respiration is used:
   a) in a bird,   b) in a plant.

9) In which three places in the body is amylase produced?

10) Where in the body is bile:
   a) produced?   b) stored?   c) used?

11) Explain why the stomach produces hydrochloric acid.

12) What is the main function of the small intestine?

13) Give two kinds of enzyme that would be useful in a biological washing powder.

14) Give an industrial use of a carbohydrase enzyme.

15) Discuss the advantages and disadvantages of using enzymes in industry.

16) Define homeostasis.

17) Write down four things that the body needs to keep fairly constant.

18) List two waste products that have to be removed from the body.

19) At what temperature do most of the enzymes in the human body work best?

20) Which area of the brain is involved in regulating the temperature of the body?

21) Write down three things that the body can do to reduce heat loss if it gets too cold.

22) What three main jobs do the kidneys do in the body?

23) Where in the body is urea produced?

24) What damage could be done in the body if the ion content is wrong?

25) Give three ways in which water is lost from the body.

26) Explain why your urine is likely to be more concentrated on a hot day.

27) Which organ monitors and controls blood glucose levels?

28) How does insulin lower the blood glucose level if it is too high?

29) How does the body respond if the blood glucose level is too low?

30) What causes diabetes? How is diabetes currently treated?

31) Describe the experiments by Banting and Best that led to the isolation of insulin.

32) Discuss the advantages and disadvantages of a pancreas transplant as a cure for diabetes.

* Answers on page 35

DNA

The first step in understanding genetics is getting to grips with DNA.

Chromosomes Are Really Long Molecules of DNA

1) DNA stands for <u>d</u>eoxyribose <u>n</u>ucleic <u>a</u>cid.
2) It contains all the <u>instructions</u> to put an organism together and <u>make it work</u>.
3) It's found in the <u>nucleus</u> of animal and plant cells, in really <u>long molecules</u> called <u>chromosomes</u>.

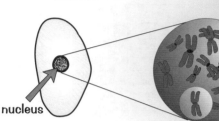

single chromosomes

a DNA molecule

nucleus

A Gene Codes for a Specific Protein

1) A <u>gene</u> is a <u>section</u> of DNA. It contains the <u>instructions</u> to make a <u>specific protein</u>.
2) Cells make <u>proteins</u> by stringing <u>amino acids</u> together in a particular order.
3) Only <u>20</u> amino acids are used, but they make up <u>thousands</u> of different <u>proteins</u>.
4) Genes simply tell cells <u>in what order</u> to put the amino acids together.
5) DNA also determines what <u>proteins</u> the cell <u>produces</u>, e.g. haemoglobin, keratin.
6) That in turn determines what <u>type of cell</u> it is, e.g. red blood cell, skin cell.

Everyone has Unique DNA...
...except identical twins and clones

Almost everyone's DNA is <u>unique</u>. The only exceptions are <u>identical twins</u>, where the two people have identical DNA, and <u>clones</u>.

<u>DNA fingerprinting</u> (or genetic fingerprinting) is a way of <u>cutting up</u> a person's DNA into small sections and then <u>separating</u> them. Every person's genetic fingerprint has a <u>unique</u> pattern (unless they're identical twins or clones of course). This means you can <u>tell people apart</u> by <u>comparing</u> <u>samples</u> of their DNA.

DNA fingerprinting is used in...

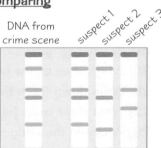

DNA from crime scene   suspect 1   suspect 2   suspect 3

1) <u>Forensic science</u> — DNA (from hair, skin flakes, blood, semen etc.) taken from a <u>crime scene</u> is compared with a DNA sample taken from a suspect. In the diagram, suspect 1's DNA has the same pattern as the DNA from the crime scene — so suspect 1 was probably at the crime scene.
2) <u>Paternity testing</u> — to see if a man is the father of a particular child.

> Some people would like there to be a national <u>genetic database</u> of everyone in the country. That way, DNA from a crime scene could be checked against <u>everyone</u> in the country to see whose it was. But others think this is a big <u>invasion of privacy</u>, and they worry about how <u>safe</u> the data would be and what <u>else</u> it might be used for. There are also <u>scientific problems</u> — <u>false positives</u> can occur if <u>errors</u> are made in the procedure or if the data is <u>misinterpreted</u>.

So the trick is — frame your twin and they'll never get you...

In the exam you might have to interpret data on <u>DNA fingerprinting for identification</u>. They'd probably give you a diagram similar to the one at the bottom of this page, and you'd have to say <u>which</u> of the <u>known</u> samples (if any) <u>matched</u> the <u>unknown</u> sample. Pretty easy — it's the two that look the same.

Cell Division — Mitosis

In order to survive and grow, our cells have got to be able to divide. And that means our DNA as well...

Mitosis Makes New Cells for Growth and Repair

Body cells normally have two copies of each chromosome — one from the organism's 'mother', and one from its 'father'. So, humans have two copies of chromosome 1, two copies of chromosome 2, etc. The diagram shows the 23 pairs of chromosomes from a human cell. The 23rd pair are a bit different — see p30.

When a body cell divides it needs to make new cells identical to the original cell — with the same number of chromosomes.

This type of cell division is called mitosis. It's used when plants and animals want to grow or to replace cells that have been damaged.

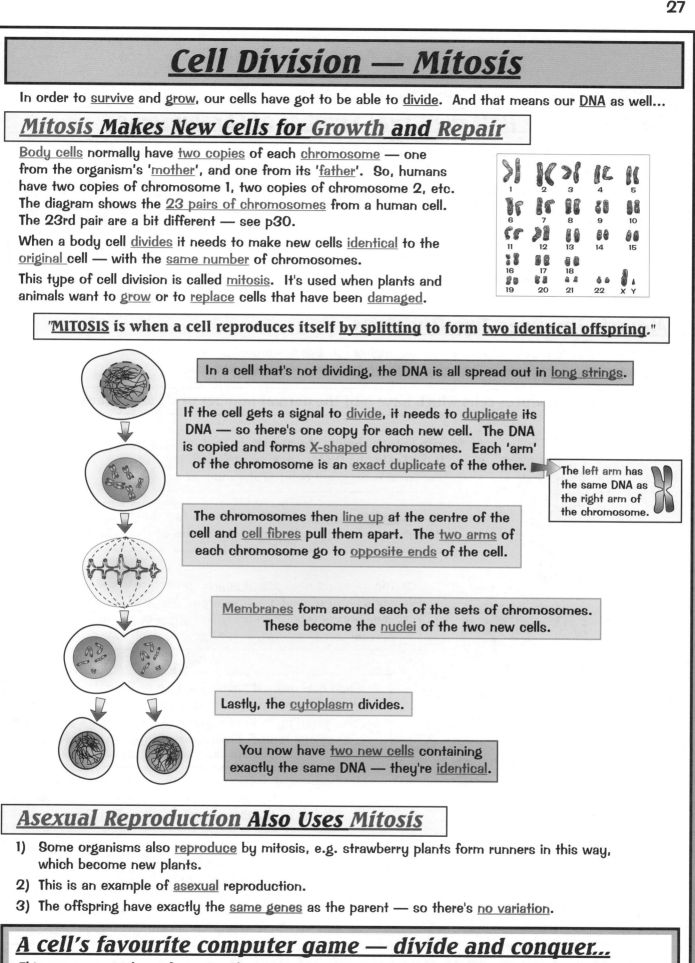

"MITOSIS is when a cell reproduces itself by splitting to form two identical offspring."

In a cell that's not dividing, the DNA is all spread out in long strings.

If the cell gets a signal to divide, it needs to duplicate its DNA — so there's one copy for each new cell. The DNA is copied and forms X-shaped chromosomes. Each 'arm' of the chromosome is an exact duplicate of the other.

The left arm has the same DNA as the right arm of the chromosome.

The chromosomes then line up at the centre of the cell and cell fibres pull them apart. The two arms of each chromosome go to opposite ends of the cell.

Membranes form around each of the sets of chromosomes. These become the nuclei of the two new cells.

Lastly, the cytoplasm divides.

You now have two new cells containing exactly the same DNA — they're identical.

Asexual Reproduction Also Uses Mitosis

1) Some organisms also reproduce by mitosis, e.g. strawberry plants form runners in this way, which become new plants.

2) This is an example of asexual reproduction.

3) The offspring have exactly the same genes as the parent — so there's no variation.

A cell's favourite computer game — divide and conquer...

This can seem tricky at first. But don't worry — just go through it slowly, one step at a time. This type of division produces identical cells, but there's another type which doesn't... (see next page)

Cell Division — Meiosis

You thought mitosis was exciting. Hah! You ain't seen nothing yet!

Gametes Have Half the Usual Number of Chromosomes

1) During sexual reproduction, two cells called gametes (sex cells) combine to form a new individual.

2) Gametes only have one copy of each chromosome. This is so that you can combine one sex cell from the 'mother' and one sex cell from the 'father' and still end up with the right number of chromosomes in body cells.

3) For example, human body cells have 46 chromosomes. The gametes have 23 chromosomes each, so that when an egg and sperm combine, you get 46 chromosomes again.

Meiosis Involves Two Divisions

To make new cells which only have half the original number of chromosomes, cells divide by meiosis. In humans, it only happens in the reproductive organs (e.g. ovaries and testes).

"MEIOSIS produces cells which have half the normal number of chromosomes."

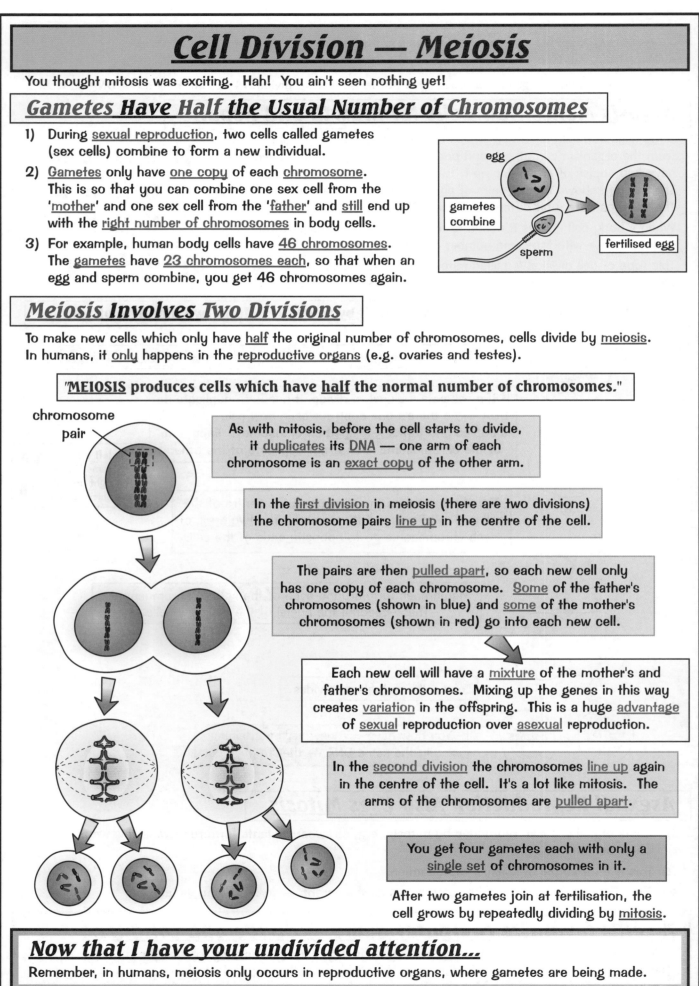

As with mitosis, before the cell starts to divide, it duplicates its DNA — one arm of each chromosome is an exact copy of the other arm.

In the first division in meiosis (there are two divisions) the chromosome pairs line up in the centre of the cell.

The pairs are then pulled apart, so each new cell only has one copy of each chromosome. Some of the father's chromosomes (shown in blue) and some of the mother's chromosomes (shown in red) go into each new cell.

Each new cell will have a mixture of the mother's and father's chromosomes. Mixing up the genes in this way creates variation in the offspring. This is a huge advantage of sexual reproduction over asexual reproduction.

In the second division the chromosomes line up again in the centre of the cell. It's a lot like mitosis. The arms of the chromosomes are pulled apart.

You get four gametes each with only a single set of chromosomes in it.

After two gametes join at fertilisation, the cell grows by repeatedly dividing by mitosis.

Now that I have your undivided attention...

Remember, in humans, meiosis only occurs in reproductive organs, where gametes are being made.

Biology 2(iii) — Genetics

Stem Cells

Stem cell research has exciting possibilities, but it's also pretty controversial.

Embryonic Stem Cells Can Turn into ANY Type of Cell

1) Most cells in your body are <u>specialised</u> for a particular job. E.g. white blood cells are brilliant at fighting invaders but can't carry oxygen, like red blood cells.

2) <u>Differentiation</u> is the process by which a cell <u>changes</u> to become <u>specialised</u> for its job. In most <u>animal</u> cells, the ability to differentiate is <u>lost</u> at an early stage, but lots of <u>plant</u> cells <u>don't</u> ever lose this ability.

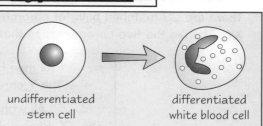

undifferentiated stem cell → differentiated white blood cell

3) Some cells are <u>undifferentiated</u>. They can develop into <u>different types of cell</u> depending on what <u>instructions</u> they're given. These cells are called <u>STEM CELLS</u>.

4) Stem cells are found in early human embryos. They're <u>exciting</u> to doctors and medical researchers because they have the potential to turn into <u>any</u> kind of cell at all. This makes sense if you think about it — <u>all</u> the <u>different types</u> of cell found in a human being have to come from those <u>few cells</u> in the early embryo.

5) Adults also have stem cells, but they're only found in certain places, like <u>bone marrow</u>. These aren't as <u>versatile</u> as embryonic stem cells — they can't turn into <u>any</u> cell type at all, only certain ones.

Stem Cells May Be Able to Cure Many Diseases

1) Medicine <u>already</u> uses adult stem cells to cure <u>disease</u>. For example, people with some <u>blood diseases</u> (e.g. <u>sickle cell anaemia</u>) can be treated by <u>bone marrow transplants</u>. Bone marrow contains <u>stem cells</u> that can turn into <u>new blood cells</u> to replace the faulty old ones.

2) Scientists can also <u>extract</u> stem cells from very early human embryos and <u>grow</u> them.

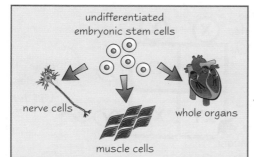

undifferentiated embryonic stem cells

nerve cells

muscle cells

whole organs

3) These embryonic stem cells could be used to <u>replace faulty cells</u> in sick people — you could make <u>beating heart muscle cells</u> for people with <u>heart disease</u>, <u>insulin-producing cells</u> for people with <u>diabetes</u>, <u>nerve cells</u> for people <u>paralysed by spinal injuries</u>, and so on.

4) To get cultures of <u>one specific type</u> of cell, researchers try to <u>control</u> the differentiation of the stem cells by changing the environment they're growing in. So far, it's still a bit hit and miss — lots more <u>research</u> is needed.

Some People Are Against Stem Cell Research

1) Some people are <u>against</u> stem cell research because they feel that human embryos <u>shouldn't</u> be used for experiments since each one is a <u>potential human life</u>.

2) Others think that curing patients who <u>already exist</u> and who are <u>suffering</u> is more important than the rights of <u>embryos</u>.

3) One fairly convincing argument in favour of this point of view is that the embryos used in the research are usually <u>unwanted ones</u> from <u>fertility clinics</u> which, if they weren't used for research, would probably just be <u>destroyed</u>. But of course, campaigners for the rights of embryos usually want this banned too.

4) These campaigners feel that scientists should concentrate more on finding and developing <u>other sources</u> of stem cells, so people could be helped <u>without</u> having to use embryos.

5) In some countries stem cell research is <u>banned</u>, but it's allowed in the UK as long as it follows <u>strict guidelines</u>.

But florists cell stems, and nobody complains about that...

The potential of stem cells is huge — but it's early days yet. Research has recently been done into getting stem cells from <u>alternative sources</u>. For example, some researchers think it might be possible to get cells from <u>umbilical cords</u> to behave like embryonic stem cells.

X and Y Chromosomes

Now for a couple of very important little chromosomes...

Your Chromosomes Control Whether You're Male or Female

There are 23 matched pairs of chromosomes in every human body cell. The 23rd pair are labelled XX or XY. They're the two chromosomes that decide whether you turn out male or female.

> All men have an X and a Y chromosome: XY
> The Y chromosome causes male characteristics.
>
> All women have two X chromosomes: XX
> The XX combination allows female characteristics to develop.

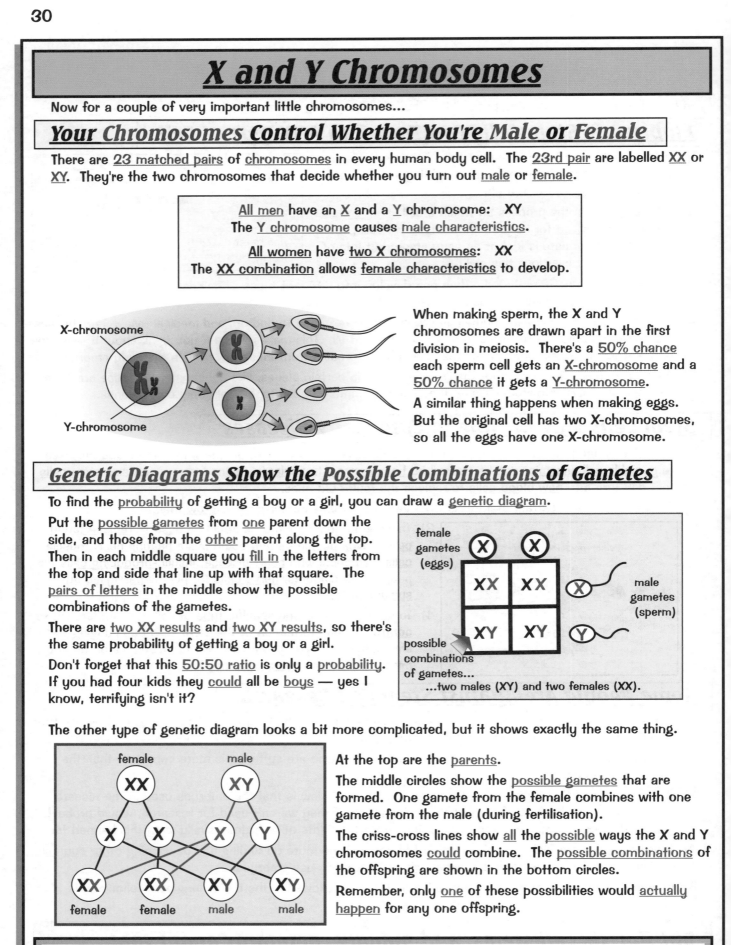

When making sperm, the X and Y chromosomes are drawn apart in the first division in meiosis. There's a 50% chance each sperm cell gets an X-chromosome and a 50% chance it gets a Y-chromosome.

A similar thing happens when making eggs. But the original cell has two X-chromosomes, so all the eggs have one X-chromosome.

Genetic Diagrams Show the Possible Combinations of Gametes

To find the probability of getting a boy or a girl, you can draw a genetic diagram.

Put the possible gametes from one parent down the side, and those from the other parent along the top. Then in each middle square you fill in the letters from the top and side that line up with that square. The pairs of letters in the middle show the possible combinations of the gametes.

There are two XX results and two XY results, so there's the same probability of getting a boy or a girl.

Don't forget that this 50:50 ratio is only a probability. If you had four kids they could all be boys — yes I know, terrifying isn't it?

female gametes (eggs)

male gametes (sperm)

possible combinations of gametes...
...two males (XY) and two females (XX).

The other type of genetic diagram looks a bit more complicated, but it shows exactly the same thing.

At the top are the parents.

The middle circles show the possible gametes that are formed. One gamete from the female combines with one gamete from the male (during fertilisation).

The criss-cross lines show all the possible ways the X and Y chromosomes could combine. The possible combinations of the offspring are shown in the bottom circles.

Remember, only one of these possibilities would actually happen for any one offspring.

Have you got the Y-factor...

Most genetic diagrams you'll see in exams concentrate on a gene, instead of a chromosome. But the principle's the same. Don't worry — there are loads of other examples on the following pages.

The Work of Mendel

Mendel Did Genetic Experiments with Pea Plants

Gregor Mendel was an Austrian monk who trained in mathematics and natural history at the University of Vienna. On his garden plot at the monastery, Mendel noted how characteristics in plants were passed on from one generation to the next.

The results of his research were published in 1866 and eventually became the foundation of modern genetics.

The diagrams show two crosses for height in pea plants that Mendel carried out...

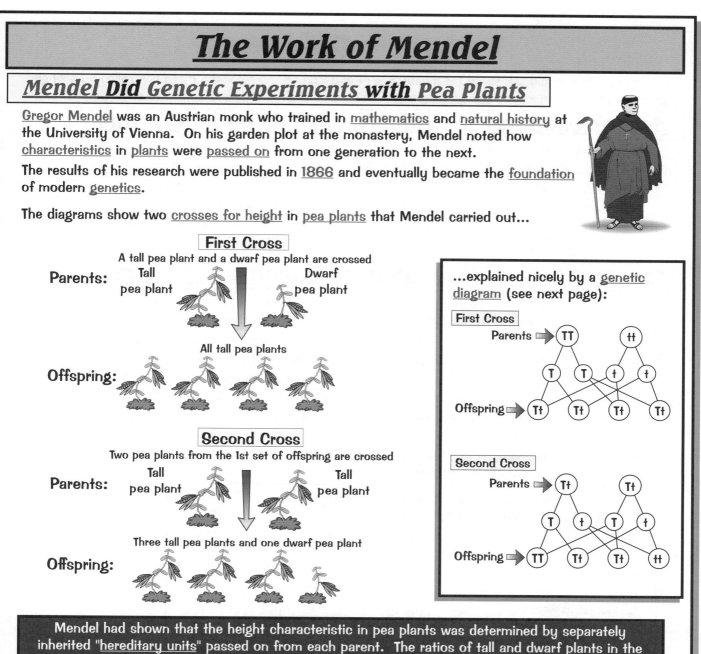

First Cross

A tall pea plant and a dwarf pea plant are crossed

Parents: Tall pea plant   Dwarf pea plant

All tall pea plants

Offspring:

...explained nicely by a genetic diagram (see next page):

First Cross

Parents ➡ TT   tt

T T t t

Offspring ➡ Tt Tt Tt Tt

Second Cross

Two pea plants from the 1st set of offspring are crossed

Parents: Tall pea plant   Tall pea plant

Three tall pea plants and one dwarf pea plant

Offspring:

Second Cross

Parents ➡ Tt   Tt

T t T t

Offspring ➡ TT Tt Tt tt

Mendel had shown that the height characteristic in pea plants was determined by separately inherited "hereditary units" passed on from each parent. The ratios of tall and dwarf plants in the offspring showed that the unit for tall plants, T, was dominant over the unit for dwarf plants, t.

Mendel Reached Three Important Conclusions

Mendel reached these three important conclusions about heredity in plants:

1) Characteristics in plants are determined by "hereditary units".
2) Hereditary units are passed on from both parents, one unit from each parent.
3) Hereditary units can be dominant or recessive — if an individual has both the dominant and the recessive unit for a characteristic, the dominant characteristic will be expressed.

We now know that the "hereditary units" are of course genes.

But in Mendel's time nobody knew anything about genes or DNA, and so the significance of his work was not to be realised until after his death.

Clearly, being a monk in the 1800s was a right laugh...

Well, there was no TV in those days, you see. Monks had to make their own entertainment. And in Mendel's case, that involved growing lots and lots of peas. He was a very clever lad, was Mendel, but unfortunately just a bit ahead of his time. Nobody had a clue what he was going on about.

Genetic Diagrams

In the exam they could ask about the inheritance of <u>any</u> kind of characteristic that's controlled by a <u>single gene</u>, because the principle's <u>always the same</u>. So here's a slightly <u>bizarre</u> example, to show you the basics.

Genetic Diagrams *Show the Possible Genes of Offspring*

1) <u>Alleles</u> are <u>different versions</u> of the <u>same gene</u>.

2) Most of the time you have <u>two copies</u> of each gene — one from each parent.

3) If they're different alleles, only one might be 'expressed' in the organism. The characteristic that appears is coded for by the <u>dominant allele</u>. The other one is <u>recessive</u>.

4) In genetic diagrams <u>letters</u> are used to represent <u>genes</u>. <u>Dominant</u> alleles are always shown with a <u>capital letter</u>, and <u>recessive</u> alleles with a <u>small letter</u>.

You Need to be Able to Interpret, Explain *and* Construct *Them*

Imagine you're cross-breeding <u>hamsters</u>, some with normal hair and a mild disposition and others with wild scratty hair and a leaning towards crazy acrobatics.

We'll use the letter 'B' (for boring) to <u>represent</u> the gene — always choose a letter whose capital looks <u>different</u> from the lower case one, so the examiners know <u>exactly</u> which one you're writing.

Let's say that the allele which causes the crazy nature is <u>recessive</u>, so we use a <u>small 'b'</u> for it, whilst normal (boring) behaviour is due to a <u>dominant allele</u>, so we represent it with a <u>capital 'B'</u>.

1) For an organism to display a <u>recessive</u> characteristic, <u>both</u> its alleles must be <u>recessive</u> — so a crazy hamster must have the alleles 'bb'.

2) However, a <u>normal hamster</u> can have <u>two possible combinations of alleles</u>, **BB** or **Bb**, because the dominant allele <u>overrules</u> the recessive one.

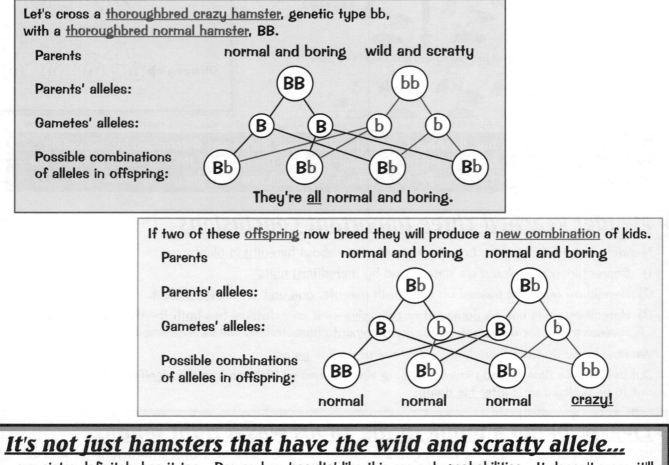

Let's cross a <u>thoroughbred crazy hamster</u>, genetic type bb, with a <u>thoroughbred normal hamster</u>, **BB**.

	normal and boring	wild and scratty
Parents		
Parents' alleles:	**BB**	**bb**
Gametes' alleles:	**B** **B**	**b** **b**
Possible combinations of alleles in offspring:	**Bb** **Bb**	**Bb** **Bb**

They're <u>all</u> normal and boring.

If two of these <u>offspring</u> now breed they will produce a <u>new combination</u> of kids.

	normal and boring	normal and boring
Parents		
Parents' alleles:	**Bb**	**Bb**
Gametes' alleles:	**B** **b**	**B** **b**
Possible combinations of alleles in offspring:	**BB** **Bb**	**Bb** **bb**
	normal normal	normal <u>crazy!</u>

It's not just hamsters that have the wild and scratty allele...

...my sister definitely has it too. Remember, '<u>results</u>' like this are only <u>probabilities</u>. It doesn't mean it'll actually happen. (Most likely, you'll end up trying to contain a mini-riot of nine lunatic baby hamsters.)

Genetic Disorders

Defective genes can cause serious problems — you need to know about two of them.

Cystic Fibrosis is Caused by a Recessive Allele

Cystic fibrosis is a genetic disorder of the cell membranes. It results in the body producing a lot of thick sticky mucus in the air passages and in the pancreas.

1) The allele which causes cystic fibrosis is a recessive allele, 'f', carried by about 1 person in 30.

2) Because it's recessive, people with only one copy of the allele won't have the disorder — they're known as carriers.

3) For a child to have the disorder, both parents must be either carriers or sufferers.

4) As the diagram shows there's a 1 in 4 chance of a child having the disorder if both parents are carriers.

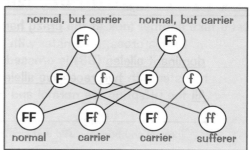

Huntington's is Caused by a Dominant Allele

Huntington's is a genetic disorder of the nervous system that's really horrible, resulting in shaking, erratic body movements and eventually severe mental deterioration.

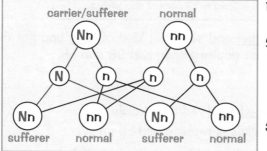

1) The disorder is caused by a dominant allele, 'N', and so can be inherited if just one parent carries the defective gene.

2) The "carrier" parent will of course be a sufferer too since the allele is dominant, but the symptoms don't start to appear until after the person is about 40. By this time the allele might already have been passed on to children and even to grandchildren.

3) As the genetic diagram shows, a person carrying the N allele has a 50% chance of passing it on to each of their children.

Embryos Can Be Screened for Genetic Disorders

1) During in vitro fertilisation (IVF), embryos are fertilised in a laboratory, and then implanted into the mother's womb. More than one egg is fertilised, so there's a better chance of the IVF being successful.

2) Before being implanted, it's possible to remove a cell from each embryo and analyse its genes.

3) Many genetic disorders could be detected in this way, such as cystic fibrosis and Huntington's.

4) Embryos with 'good' genes would be implanted into the mother — the ones with 'bad' genes destroyed.

There is a huge debate raging about embryonic screening. Here are some arguments for and against it.

Against Embryonic Screening	For Embryonic Screening
1) There may come a point where everyone wants to screen their embryos so they can pick the most 'desirable' one, e.g. they want a blue eyed, blonde haired, intelligent boy.	1) It will help to stop people suffering.
	2) There are laws to stop it going too far. At the moment parents cannot even select the sex of their baby (unless it's for health reasons).
2) The rejected embryos are destroyed — they could have developed into humans.	3) During IVF, most of the embryos are destroyed anyway — screening just allows the selected one to be healthy.
3) It implies that people with genetic problems are 'undesirable' — this could increase prejudice.	4) Treating disorders costs the Government (and the taxpayers) a lot of money.

Embryonic screening — it's a tricky one...

There's a nice moral argument for you to consider on this page. In the exam you may be asked your opinion — make sure you can back it up with good reasons, and consider other points of view.

More Genetic Diagrams

You've got to be able to <u>predict</u> and <u>explain</u> the outcomes of crosses between individuals for each <u>possible</u> <u>combination</u> of <u>dominant</u> and <u>recessive alleles</u> of a gene. If you've got your head round all this genetics lark you should be able to draw a <u>genetic diagram</u> and <u>work it out</u> — but it'll make it easier if you've seen them all before. So here are some examples for you:

All the Offspring are Normal

Let's take another look at the <u>crazy hamster</u> example:

In this cross, a hamster with <u>two</u> <u>dominant alleles</u> (BB) is crossed with a hamster with <u>two recessive alleles</u> (bb). <u>All</u> the offspring are normal and boring.

But, if you crossed a hamster with <u>two</u> <u>dominant alleles</u> (BB) with a hamster with <u>a</u> <u>dominant</u> and <u>a recessive allele</u> (Bb), you would also get <u>all</u> normal and boring offspring.

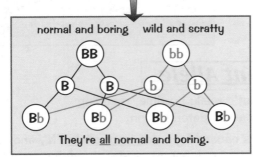

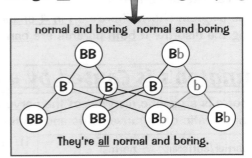

To find out <u>which</u> it was you'd have to <u>breed the offspring together</u> and see what kind of <u>ratio</u> you got that time — then you'd have a good idea. If it was <u>3:1</u>, it's likely that you originally had **BB** and **bb**.

There's a 3:1 Ratio in the Offspring

<u>Sickle cell anaemia</u> is a genetic disorder characterised by <u>funny-shaped</u> red blood cells.

It's caused by a <u>recessive</u> allele 'a' (for anaemia). The normal allele is represented by an 'A'.

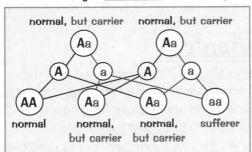

If two people who <u>carry</u> the sickle cell anaemia allele have children, the <u>probability</u> of each child suffering from the disorder is 1 in 4 — <u>25%</u>.

The ratio you'd expect in the children is <u>3:1</u>, non-sufferer:sufferer.

If you see this ratio in the offspring you know <u>both</u> parents must have the <u>two different alleles</u>.

Be careful with this one — it may be disguised as a <u>1:2:1</u> ratio (normal:carrier:sufferer), but it means the same thing.

There's a 1:1 Ratio in the Offspring

A cat with <u>long hair</u> was bred with another cat with <u>short hair</u>. The long hair is caused by a <u>dominant</u> allele 'H', and the short hair by a <u>recessive</u> allele 'h'.

They had 8 kittens — 4 with long hair and 4 with short hair.

This is a <u>1:1</u> ratio — it's what you'd expect when a parent with only <u>one dominant allele</u> (Hh) is crossed with a parent with <u>two recessive alleles</u> (hh).

It's enough to make you go cross-eyed...

Remember that these are <u>only probabilities</u>, so you need <u>loads</u> of organisms in each generation to see a <u>reliable ratio</u>. That's why people tend to do genetic experiments with quick-breeding organisms like <u>fruit</u> <u>flies</u>. And of course it still won't be an <u>exact</u> ratio (just because of normal chance) — you might get 69 normal fruit flies and 31 crazy fruit flies out of 100. Not exact, but close enough to a 3:1 ratio.

Revision Summary for Biology 2(iii)

This section contains loads of new stuff all about genetics — it's even got a couple of moral dilemmas for you to ponder over. Your DNA contains all the instructions needed to make you. But your DNA can only account for so much, whether you can roll your tongue for example. You can't blame it for what colour you decide to dye your hair, or for leaving your smelly socks on the stairs though...

Use these questions to find out what you know about it all — and what you don't. Then look back and learn the bits you don't know. Then try the questions again, and again...

1) What is a gene?
2) Explain how DNA controls the activities of a cell.
3) Explain how DNA fingerprinting is used in forensic science.
4) Some people would like there to be a genetic database of everyone in the country. Discuss the advantages and disadvantages of such a database for use in forensic science.
5) What is mitosis used for in the human body? Describe the four steps in mitosis.
6) Name the other type of cell division, and say where it happens in the body of a human male.
7) Explain why sexual reproduction produces more variation than asexual reproduction.
8) What type of cell division does a fertilised egg use to grow into a new organism?
9) What is differentiation in a cell?
10) Give an example of a tissue where stem cells are found in adults.
11) Why are embryonic stem cells currently thought to be more useful than adult stem cells?
12) Give three ways that embryonic stem cells could be used to cure diseases.
13) Discuss the moral arguments for and against embryonic stem cell research in the UK.
14) Which chromosome in the human body causes male characteristics?
15) Copy and complete the diagrams below to show what happens to the X and Y chromosomes during reproduction.

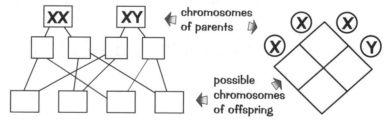

16)* A couple have three boys already. What is the probability that their fourth child will be a girl? (Hint: this may be a nasty trick question — don't be fooled.)
17) List three important conclusions that Mendel reached following his experiments with pea plants.
18) What were the "hereditary units" that Mendel concluded were controlling the characteristics of his pea plants?
19) The significance of Mendel's work was not realised until 1900, 16 years after Mendel died. Suggest why the importance of the work wasn't understood at the time.
20) What is an allele?
21) Cystic fibrosis is caused by a recessive allele. If both parents are carriers, what is the probability of their child: a) being a carrier,   b) suffering from the disorder?
22) During in vitro fertilisation, it is possible to screen embryos for various genetic disorders before they're implanted into the mother. Only the "good" embryos would be chosen for implantation. Summarise the main arguments for and against embryonic screening.
23)* Blue colour in a plant is carried on a recessive allele, b. The dominant allele, B, gives white flowers. In the first generation after a cross, all the flowers are white. These are bred together and the result is a ratio of 54 white : 19 blue. What were the alleles of the flowers used in the first cross?

*Answers at the bottom of the page

Biology Answers

Biology 2(ii) P25 Revn Summary: 5(a) 1.65 b) stomach
Biology 2(iii) P35 Revn Summary: 16) 50% 23) BB and bb.

Biology 2(iii) — Genetics

Atoms

There are quite a few different (and equally useful) models of the atom — but chemists tend to like this nuclear model best. You can use it to explain pretty much the whole of Chemistry... which is nice.

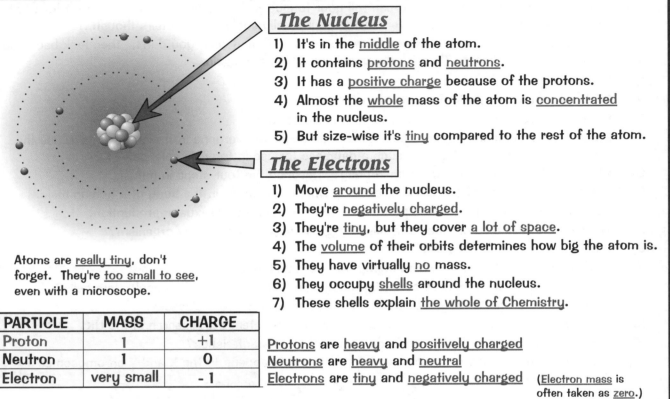

The Nucleus

1) It's in the middle of the atom.
2) It contains protons and neutrons.
3) It has a positive charge because of the protons.
4) Almost the whole mass of the atom is concentrated in the nucleus.
5) But size-wise it's tiny compared to the rest of the atom.

The Electrons

1) Move around the nucleus.
2) They're negatively charged.
3) They're tiny, but they cover a lot of space.
4) The volume of their orbits determines how big the atom is.
5) They have virtually no mass.
6) They occupy shells around the nucleus.
7) These shells explain the whole of Chemistry.

Atoms are really tiny, don't forget. They're too small to see, even with a microscope.

PARTICLE	MASS	CHARGE
Proton	1	+1
Neutron	1	0
Electron	very small	-1

Protons are heavy and positively charged
Neutrons are heavy and neutral
Electrons are tiny and negatively charged (Electron mass is often taken as zero.)

Number of Protons Equals Number of Electrons

1) Neutral atoms have no charge overall.
2) The charge on the electrons is the same size as the charge on the protons — but opposite.
3) This means the number of protons always equals the number of electrons in a neutral atom.
4) If some electrons are added or removed, the atom becomes charged and is then an ion.

Atomic Number and Mass Number Describe an Atom

These two numbers tell you how many of each kind of particle an atom has.

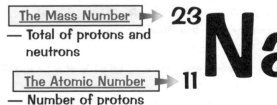

The Mass Number
— Total of protons and neutrons

The Atomic Number
— Number of protons

1) The atomic (proton) number tells you how many protons there are.
2) Atoms of the same element all have the same number of protons — so atoms of different elements will have different numbers of protons.

3) To get the number of neutrons, just subtract the atomic number from the mass number.
4) The mass (nucleon) number is always the biggest number.
   On a periodic table the mass number is actually the relative atomic mass.
5) The mass number tends to be roughly double the proton number.
6) Which means there's about the same number of protons as neutrons in any nucleus.

Number of protons = number of electrons...

This stuff might seem a bit useless at first, but it should be permanently engraved into your mind. If you don't know these basic facts, you've got no chance of understanding the rest of Chemistry.

Elements, Compounds and Isotopes

Elements Consist of One Type of Atom Only

Quite a lot of everyday substances are elements:

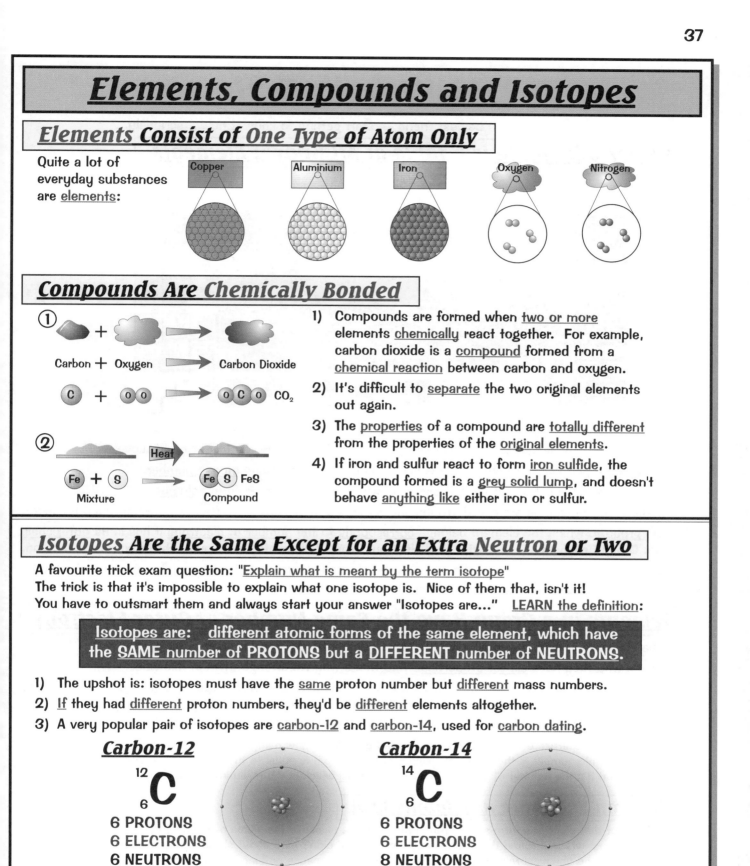

Copper Aluminium Iron Oxygen Nitrogen

Compounds Are Chemically Bonded

Carbon + Oxygen ⟶ Carbon Dioxide

C + O O ⟶ O C O CO₂

Heat

Fe + S ⟶ Fe S FeS
Mixture   Compound

1) Compounds are formed when <u>two or more</u> elements <u>chemically</u> react together. For example, carbon dioxide is a <u>compound</u> formed from a <u>chemical reaction</u> between carbon and oxygen.

2) It's difficult to <u>separate</u> the two original elements out again.

3) The <u>properties</u> of a compound are <u>totally different</u> from the properties of the <u>original elements</u>.

4) If iron and sulfur react to form <u>iron sulfide</u>, the compound formed is a <u>grey solid lump</u>, and doesn't behave <u>anything like</u> either iron or sulfur.

Isotopes Are the Same Except for an Extra Neutron or Two

A favourite trick exam question: "<u>Explain what is meant by the term isotope</u>"
The trick is that it's impossible to explain what one isotope is. Nice of them that, isn't it!
You have to outsmart them and always start your answer "Isotopes are..."   <u>LEARN</u> the definition:

> Isotopes are: <u>different atomic forms</u> of the <u>same element</u>, which have the <u>SAME</u> number of <u>PROTONS</u> but a <u>DIFFERENT</u> number of <u>NEUTRONS</u>.

1) The upshot is: isotopes must have the <u>same</u> proton number but <u>different</u> mass numbers.

2) <u>If</u> they had <u>different</u> proton numbers, they'd be <u>different</u> elements altogether.

3) A very popular pair of isotopes are <u>carbon-12</u> and <u>carbon-14</u>, used for <u>carbon dating</u>.

Carbon-12
$^{12}_{6}C$
6 PROTONS
6 ELECTRONS
6 NEUTRONS

Carbon-14
$^{14}_{6}C$
6 PROTONS
6 ELECTRONS
8 NEUTRONS

The <u>number</u> of electrons decides the <u>chemistry</u> of the element. If the <u>proton number</u> is the same (that is, the <u>number of protons</u> is the same) then the <u>number of electrons</u> must be the same, so the <u>chemistry</u> is the same. The <u>different</u> number of <u>neutrons</u> in the nucleus <u>doesn't</u> affect the chemical behaviour <u>at all</u>.

Will this be in your exam — isotope so...

<u>Carbon-14</u> is unstable. It makes up about one ten-millionth of the carbon in living things. When things die, the C-14 is trapped inside the dead material, and it gradually <u>decays</u> into nitrogen. So by measuring the proportion of C-14 found in some old axe handle you can calculate <u>how long ago</u> it was <u>living</u> wood.

Chemistry 2(i) — Bonding and Reactions

The Periodic Table

The periodic table is a chemist's bestest friend — start getting to know it now... seriously...

The Periodic Table is a Table of All Known Elements

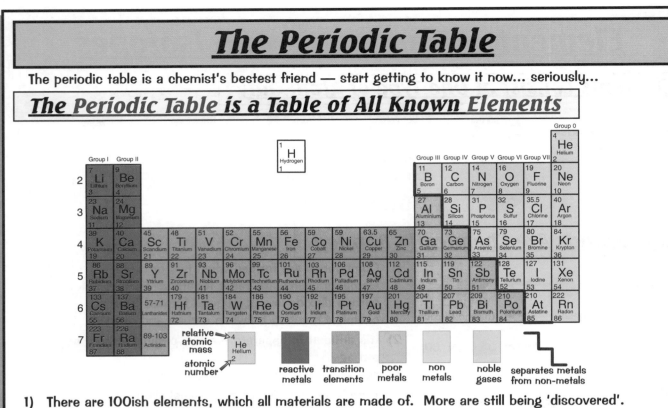

1) There are 100ish elements, which all materials are made of. More are still being 'discovered'.

2) The <u>modern</u> periodic table shows the elements in order of ascending <u>atomic number</u>.

3) The periodic table is laid out so that elements with <u>similar properties</u> form <u>columns</u>.

4) These <u>vertical columns</u> are called <u>groups</u> and roman numerals are often (but not always) used for them.

5) The <u>group</u> to which the element belongs <u>corresponds</u> to the <u>number of electrons</u> it has in its <u>outer shell</u>. E.g. Group 1 elements have 1 outer shell electron, Group 7 elements have 7 outer shell electrons and so on.

6) Some of the groups have special names. <u>Group 1</u> elements are called <u>alkali metals</u>. <u>Group 7</u> elements are called <u>halogens</u>, and <u>Group 0</u> are called the <u>noble gases</u>.

Elements in a Group Have the Same Number of Outer Electrons

1) The elements in each <u>group</u> all have the same number of <u>electrons</u> in their <u>outer shell</u>.

2) That's why they have <u>similar properties</u>. And that's why we arrange them in this way.

3) When only a handful of elements were known, the periodic table was made by looking at the properties of the elements and arranging them in groups — the same groups that they are in today.

4) This idea is extremely important to chemistry — so make sure you understand it.

> The properties of the elements are decided <u>entirely</u> by how many electrons they have.
> Atomic number is therefore very significant because it is equal to the number of electrons each atom has.
> But it's the number of electrons in the <u>outer shell</u> which is the really important thing.

Electron Shells are just Totally Brill

The fact that electrons form shells around atoms is the basis for the whole of Chemistry.
If they just whizzed round the nucleus any old how and didn't care about shells or any of that stuff there'd be no chemical reactions. No nothing in fact — because nothing would happen.
The atoms would just slob about, all day long. Just like teenagers.
But amazingly, they do form shells (if they didn't, we wouldn't even be here to wonder about it), and the electron arrangement of each atom determines the whole of its chemical behaviour.
Phew. I mean electron arrangements explain practically the whole Universe. They're just totally brill.

I've got a periodic table — Queen Anne legs and everything...

Physicists can produce <u>new</u> elements in particle accelerators, but they're all <u>radioactive</u>. Most only last a fraction of a second before they decay. They haven't even got round to giving most of them proper names yet, but then even "element 114" sounds pretty cool when you say it in Latin — <u>ununquadium</u>...

Electron Shells

The fact that electrons occupy "shells" around the nucleus is what causes the whole of chemistry.
Remember that, and watch how it applies to each bit of it. It's ace.

Electron Shell Rules:

1) Electrons always occupy <u>shells</u> (sometimes called <u>energy levels</u>).

2) The <u>lowest</u> energy levels are <u>always filled first</u> — these are the ones closest to the nucleus.

3) Only <u>a certain number</u> of electrons are allowed in each shell:
<u>1st shell:</u> 2   <u>2nd Shell:</u> 8   <u>3rd Shell:</u> 8

4) Atoms are much <u>happier</u> when they have <u>full electron shells</u> — like the <u>noble gases</u> in <u>Group 0</u>.

5) In most atoms the <u>outer shell</u> is <u>not full</u> and this makes the atom want to <u>react</u>.

3rd
2nd
1st

3rd shell still filling

Follow the Rules to Work Out Electron Configurations

You need to know the <u>electron configurations</u> for the first <u>20</u> elements (things get a bit more complicated after that). But they're not hard to work out. For a quick example, take nitrogen. <u>Follow the steps...</u>

1) The periodic table tells us nitrogen has <u>seven</u> protons... so it must have <u>seven</u> electrons.

2) Follow the 'Electron Shell Rules' above. The <u>first</u> shell can only take 2 electrons and the <u>second</u> shell can take a <u>maximum</u> of 8 electrons.

3) So the electron configuration for nitrogen <u>must</u> be <u>2, 5</u>. Easy peasy.

4) Now <u>you</u> try it for argon.

The periodic table has a big gap here where the transition metals fit in on row four.

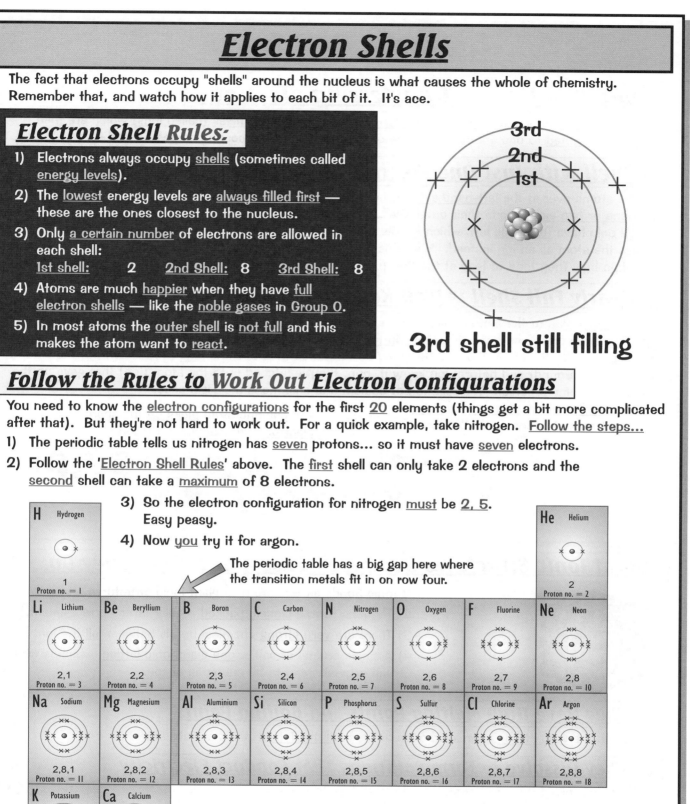

Answer... To calculate the electron configuration of argon, <u>follow the rules</u>. It's got 18 protons, so it <u>must</u> have 18 electrons. The first shell must have <u>2</u> electrons, the second shell must have <u>8</u>, and so the third shell must have <u>8</u> as well. It's as easy as <u>2, 8, 8</u>.

One little duck and two fat ladies — 2, 8, 8...

You need to know enough about electron shells to draw out that <u>whole diagram</u> at the bottom of the page without looking at it. Obviously, you don't have to learn each element separately, just <u>learn the pattern</u>. Cover the page: using a periodic table, find the atom with the electron configuration 2, 8, 6.

Ionic Bonding

Ionic Bonding — Transferring Electrons

In ionic bonding, atoms lose or gain electrons to form charged particles (called ions) which are then strongly attracted to one another (because of the attraction of opposite charges, + and –).

A Shell with Just One Electron is Well Keen to Get Rid...

All the atoms over at the left-hand side of the periodic table, e.g. sodium, potassium, calcium etc. have just one or two electrons in their outer shell. And they're pretty keen to get shot of them, because then they'll only have full shells left, which is how they like it. So given half a chance they do get rid, and that leaves the atom as an ion instead. Now ions aren't the kind of things that sit around quietly watching the world go by. They tend to leap at the first passing ion with an opposite charge and stick to it like glue.

A Nearly Full Shell is Well Keen to Get That Extra Electron...

On the other side of the periodic table, the elements in Group 6 and Group 7, such as oxygen and chlorine, have outer shells which are nearly full. They're obviously pretty keen to gain that extra one or two electrons to fill the shell up. When they do of course they become ions (you know, not the kind of things to sit around) and before you know it, pop, they've latched onto the atom (ion) that gave up the electron a moment earlier. The reaction of sodium and chlorine is a classic case:

The sodium atom gives up its outer electron and becomes an Na+ ion.

The chlorine atom picks up the spare electron and becomes a Cl⁻ ion.

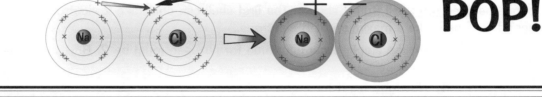

POP!

Giant Ionic Structures Don't Melt Easily, but When They Do...

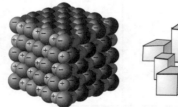

1) Ionic bonds always produce giant ionic structures.
2) The ions form a closely packed regular lattice arrangement.
3) There are very strong chemical bonds between all the ions.
4) A single crystal of salt is one giant ionic lattice, which is why salt crystals tend to be cuboid in shape.

1) They Have High Melting Points and Boiling Points

due to the very strong chemical bonds between all the ions in the giant structure.

2) They Dissolve to Form Solutions That Conduct Electricity

When dissolved the ions separate and are all free to move in the solution, so obviously they'll carry electric current. Dissolved lithium salts are used to make rechargeable batteries.

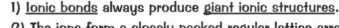
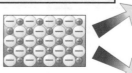

Dissolved in Water

3) They Conduct Electricity When Molten

When it melts, the ions are free to move and they'll carry electric current.

Melted

Giant ionic lattices — all over your chips...

Because they conduct electricity when they're dissolved in water, ionic compounds are used to make some types of battery. In the olden days, most batteries had actual liquid in, so they tended to leak all over the place. Now they've come up with a sort of paste that doesn't leak but still conducts. Clever.

Electron Shells and Ions

Groups 1 & 2 and 6 & 7 are the Most Likely to Form Ions

1) Remember, atoms that have <u>lost</u> or <u>gained</u> an electron (or electrons) are <u>ions</u>.
2) The elements that most readily form ions are those in Groups 1, 2, 6 and 7.
3) <u>Group 1 and 2 elements</u> are <u>metals</u> and they <u>lose</u> electrons to form <u>+ve ions</u> or <u>cations</u>.
4) <u>Group 6 and 7 elements</u> are <u>non-metals</u>. They <u>gain</u> electrons to form <u>–ve ions</u> or <u>anions</u>.
5) Make sure you know these easy ones:

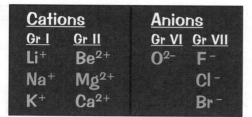

Cations		Anions	
<u>Gr I</u>	<u>Gr II</u>	<u>Gr VI</u>	<u>Gr VII</u>
Li^+	Be^{2+}	O^{2-}	F^-
Na^+	Mg^{2+}		Cl^-
K^+	Ca^{2+}		Br^-

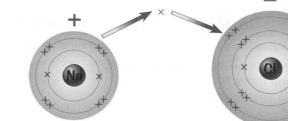

6) When any of the above cations <u>react</u> with the anions, they form <u>ionic bonds</u>.
7) Only elements at <u>opposite sides</u> of the periodic table will form ionic bonds, e.g. Na and Cl, where one of them becomes a <u>cation</u> (+ve) and one becomes an <u>anion</u> (–ve).

> Remember, the + and – charges we talk about, e.g. Na^+ for sodium, just tell you <u>what type of ion the atom WILL FORM</u> in a chemical reaction. In sodium <u>metal</u> there are <u>only neutral sodium atoms, Na</u>. The Na^+ ions <u>will only appear</u> if the sodium metal <u>reacts</u> with something like water or chlorine.

Show the Electronic Structure of Simple Ions With Brackets [ ]

A useful way of representing ions is by specifying the <u>ion's name</u>, followed by its <u>electron configuration</u> and the <u>charge</u> on the ion. For example, the electronic structure of the sodium ion Na^+ can be represented by Na [2, 8]$^+$. That's the electron configuration followed by the charge on the ion. Simple enough. A few <u>ions</u> and the <u>ionic compounds</u> they form are shown below.

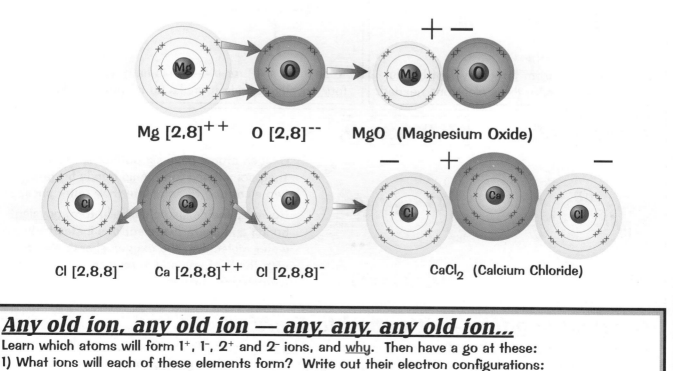

Mg [2,8]$^{++}$ O [2,8]$^{--}$ MgO (Magnesium Oxide)

Cl [2,8,8]$^-$ Ca [2,8,8]$^{++}$ Cl [2,8,8]$^-$ CaCl$_2$ (Calcium Chloride)

Any old ion, any old ion — any, any, any old ion...

Learn which atoms will form 1$^+$, 1$^-$, 2$^+$ and 2$^-$ ions, and <u>why</u>. Then have a go at these:
1) What ions will each of these elements form? Write out their electron configurations:
  a) potassium, b) aluminium, c) beryllium, d) sulfur, e) fluorine (using a periodic table)   Answers on page 71

Covalent Bonding

Covalent Bonds — Sharing Electrons

1) Sometimes atoms prefer to make covalent bonds by sharing electrons with other atoms.
2) This way both atoms feel that they have a full outer shell, and that makes them happy.
3) Each covalent bond provides one extra shared electron for each atom.
4) Each atom involved has to make enough covalent bonds to fill up its outer shell.
5) Learn these seven important examples:

You only have to draw the outer shell of electrons.

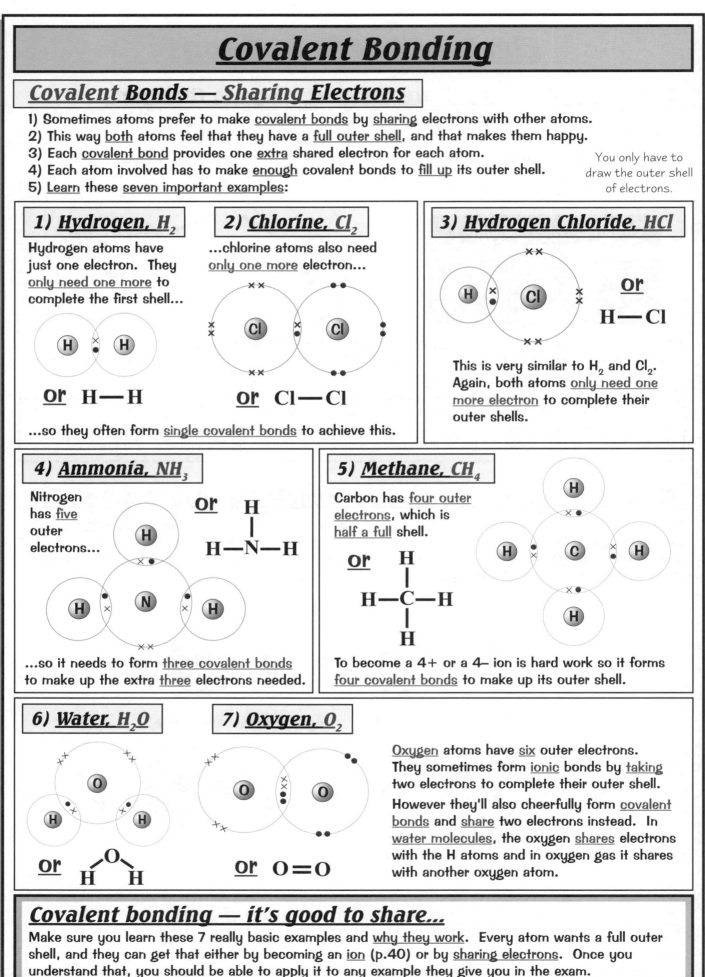

1) Hydrogen, H_2

Hydrogen atoms have just one electron. They only need one more to complete the first shell...

or H — H

...so they often form single covalent bonds to achieve this.

2) Chlorine, Cl_2

...chlorine atoms also need only one more electron...

or Cl — Cl

3) Hydrogen Chloride, HCl

or H — Cl

This is very similar to H_2 and Cl_2. Again, both atoms only need one more electron to complete their outer shells.

4) Ammonia, NH_3

Nitrogen has five outer electrons...

or H — N — H (with H above N)

...so it needs to form three covalent bonds to make up the extra three electrons needed.

5) Methane, CH_4

Carbon has four outer electrons, which is half a full shell.

or H — C — H (with H above and below C)

To become a 4+ or a 4− ion is hard work so it forms four covalent bonds to make up its outer shell.

6) Water, H_2O

or H — O — H (O above)

7) Oxygen, O_2

or O = O

Oxygen atoms have six outer electrons. They sometimes form ionic bonds by taking two electrons to complete their outer shell.

However they'll also cheerfully form covalent bonds and share two electrons instead. In water molecules, the oxygen shares electrons with the H atoms and in oxygen gas it shares with another oxygen atom.

Covalent bonding — it's good to share...

Make sure you learn these 7 really basic examples and why they work. Every atom wants a full outer shell, and they can get that either by becoming an ion (p.40) or by sharing electrons. Once you understand that, you should be able to apply it to any example they give you in the exam.

Covalent Substances: Two Kinds

Substances formed from covalent bonds can either be simple molecules or giant structures.

Simple Molecular Substances

1) The atoms form very strong covalent bonds to form small molecules of several atoms.
2) By contrast, the forces of attraction between these molecules are very weak.
3) The result of these feeble intermolecular forces is that the melting and boiling points are very low, because the molecules are easily parted from each other.
4) Most molecular substances are gases or liquids at room temperature.
5) Molecular substances don't conduct electricity, simply because there are no ions.
6) You can usually tell a molecular substance just from its physical state, which is always kinda 'mushy' — i.e. liquid or gas or an easily-melted solid.

Very weak inter-molecular forces

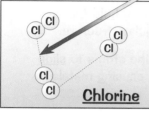
Chlorine

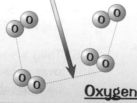
Oxygen

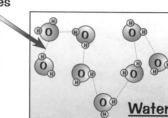
Water

Giant Covalent Structures

1) These are similar to giant ionic structures except that there are no charged ions.
2) All the atoms are bonded to each other by strong covalent bonds.
3) They have very high melting and boiling points.
4) They don't conduct electricity — not even when molten.
5) They're usually insoluble in water.
6) The main examples are diamond and graphite, which are both made only from carbon atoms.

Diamond

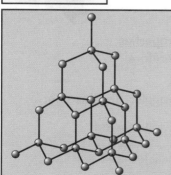

Each carbon atom forms four covalent bonds in a very rigid giant covalent structure. This structure makes diamond the hardest natural substance, so it's used for drill tips. And it's all pretty and sparkly too.

Graphite

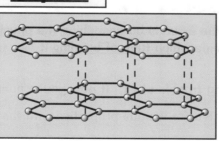

Each carbon atom only forms three covalent bonds, creating layers which are free to slide over each other. The layers are held together so loosely that they can be rubbed off onto paper — that's how a pencil works. It also leaves free electrons, so graphite is the only non-metal which is a good conductor of electricity.

Silicon Dioxide (Silica)

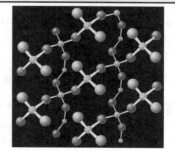

Sometimes called silica, this is what sand is made of.
Each grain of sand is one giant structure of silicon and oxygen. Silica can be melted down with sodium carbonate (Na_2CO_3) and limestone ($CaCO_3$) to make glass.

Carbon is a girl's best friend...

The two different types of covalent substance are very different, make sure you know about them both.

Metallic Structures

Metal Properties Are All Due to the Sea of Free Electrons

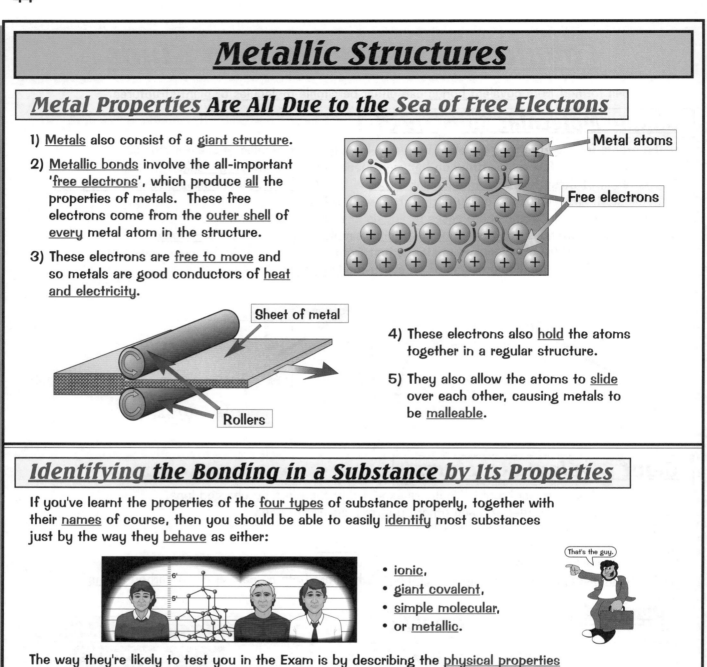

1) <u>Metals</u> also consist of a <u>giant structure</u>.

2) <u>Metallic bonds</u> involve the all-important '<u>free electrons</u>', which produce <u>all</u> the properties of metals. These free electrons come from the <u>outer shell</u> of <u>every</u> metal atom in the structure.

3) These electrons are <u>free to move</u> and so metals are good conductors of <u>heat and electricity</u>.

Metal atoms

Free electrons

Sheet of metal

Rollers

4) These electrons also <u>hold</u> the atoms together in a regular structure.

5) They also allow the atoms to <u>slide</u> over each other, causing metals to be <u>malleable</u>.

Identifying the Bonding in a Substance by Its Properties

If you've learnt the properties of the <u>four types</u> of substance properly, together with their <u>names</u> of course, then you should be able to easily <u>identify</u> most substances just by the way they <u>behave</u> as either:

That's the guy.

- <u>ionic</u>,
- <u>giant covalent</u>,
- <u>simple molecular</u>,
- or <u>metallic</u>.

The way they're likely to test you in the Exam is by describing the <u>physical properties</u> of a substance and asking you to decide <u>which type of bonding</u> it has and therefore what type of material it is.

If you know your onions you'll have no trouble at all. If not, you're gonna struggle. Try this one:

<u>Example</u>: Four substances were tested for various properties with the following results:

Substance	Melting point (°C)	Boiling point (°C)	Good electrical conductor?
A	−218.4	−182.96	No
B	1535	2750	Yes
C	1410	2355	No
D	801	1413	When molten

Identify the type of bonding in each substance. (Answers at the bottom of page 71.)

Right — back to page 40 and go through all that again...

You have to be able to identify the bonding in <u>any</u> substance based on its properties — and explain <u>why</u>. Make sure you can answer a question the other way round as well, e.g. "Fireclay is made from silica and other giant molecular substances. Explain why it might be suitable for lining a kiln."

New Materials

New materials are continually being developed, with new properties. The two groups of materials you really need to know about are <u>smart materials</u> and <u>nanomaterials</u>.

Smart Materials Have Some Really Weird Properties

1) <u>Smart</u> materials <u>behave differently</u> depending on the <u>conditions</u>, e.g. temperature.

2) A good example is <u>nitinol</u> — a "<u>shape memory alloy</u>".

   It's a metal, but when it's cool you can <u>bend it</u> and <u>twist it</u> like rubber. Bend it too far, though, and it stays bent. But here's the really clever bit — if you heat it above a certain temperature, it goes back to a "<u>remembered</u>" shape (hence the name). It's really handy for glasses frames. If you accidentally bend them, you can just pop them into a bowl of hot water and they'll <u>jump</u> back <u>into shape</u>.
   Nitinol is made from about half <u>nickel</u>, half <u>titanium</u>.

3) Other examples of smart materials include <u>dyes</u> that change <u>colour</u> depending on <u>temperature</u> or <u>light intensity</u>, <u>liquids</u> that turn <u>solid</u> when you put them in a <u>magnetic field</u>, and materials that <u>expand</u> or <u>contract</u> when you put an <u>electric current</u> through them.

Nanomaterials Are Really Really Really Really Tiny   ...smaller than that.

1) Really tiny particles, <u>1–100 nanometres</u> across, are called 'nanoparticles' (1 nm = 0.000 000 001 m).

2) Nanoparticles include <u>fullerenes</u>. These are molecules of <u>carbon</u>, shaped like <u>hollow balls</u> or <u>closed tubes</u>. Each carbon atom forms <u>three</u> covalent bonds with its neighbours, leaving <u>free electrons</u> that can <u>conduct</u> electricity.

3) The smallest fullerene is <u>buckminster fullerene</u>, which has <u>sixty</u> carbon atoms joined in a <u>ball</u> — its molecular formula is C_{60}.

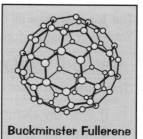

Buckminster Fullerene

Fullerenes can be joined together to form <u>nanotubes</u> — teeny tiny hollow carbon tubes, a few nanometres across:

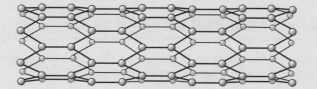

A nanoparticle has very <u>different properties</u> from the 'bulk' chemical that it's made from — e.g. <u>fullerenes</u> have different properties from big <u>lumps of carbon</u>.

a) All those covalent bonds make carbon nanotubes <u>very strong</u>. They can be used to reinforce graphite in tennis rackets and to make stronger, lighter building materials.

b) Nanotubes <u>conduct</u> electricity, so they can be used in tiny <u>electric circuits</u> for computer chips.

4) Nanoparticles have some <u>very useful properties</u>:

a) They have a <u>huge surface area</u>, so they could help make great industrial <u>catalysts</u> (see page 59) — individual catalyst molecules could be attached to carbon nanotubes.

b) With nanoparticles, you can build surfaces with very <u>specific properties</u>. That means you can use them to make <u>sensors</u> to detect one type of molecule and nothing else. These <u>highly specific</u> sensors are already being used to test water purity.

5) Nanoparticles can be made by <u>molecular engineering</u> but this is <u>really hard</u>. Molecular engineering is building a product <u>molecule-by-molecule</u> to a specific design — either by positioning each molecule exactly where you want it or by starting with a bigger structure and taking bits off it.

Bendy specs, tennis rackets and computer chips — cool...

Some nanoparticles have really <u>unexpected properties</u>. Silver's normally very unreactive, but silver nanoparticles can kill bacteria. Gold nanoparticles aren't gold-coloured — they're either red or purple.

Balancing Equations

Equations need a lot of practice if you're going to get them right.
Every time you do an equation you need to practise getting it right rather than skating over it.

The Symbol Equation Shows the Atoms on Both Sides:

A chemical reaction can be described by the process reactants → products.

e.g.   magnesium reacts with oxygen to produce magnesium oxide.

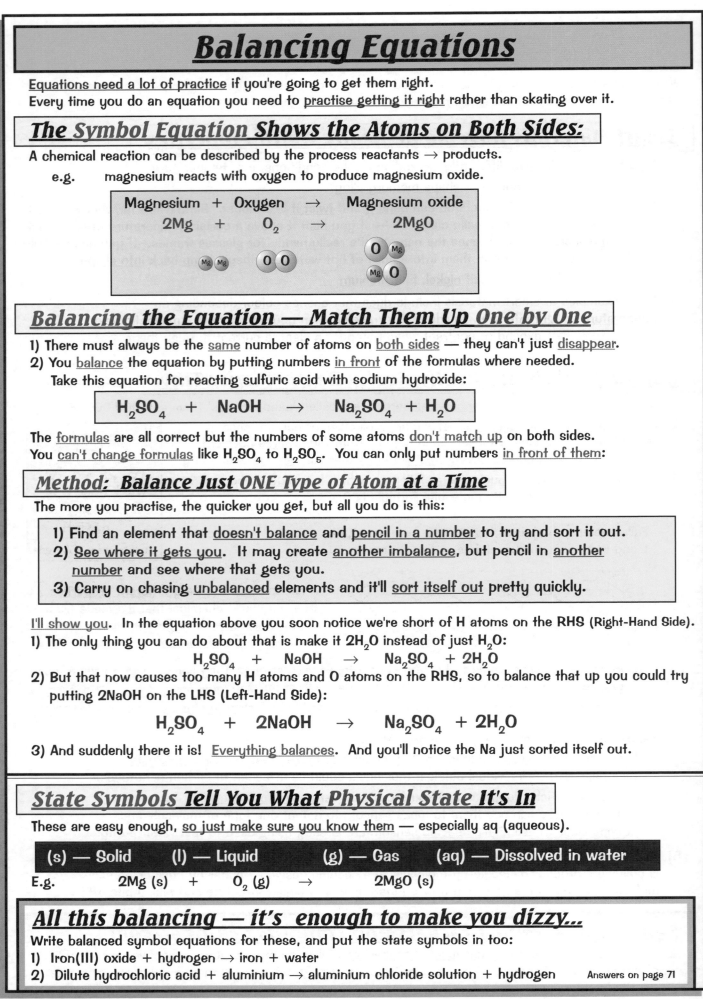

$$\text{Magnesium} + \text{Oxygen} \rightarrow \text{Magnesium oxide}$$
$$2Mg + O_2 \rightarrow 2MgO$$

Balancing the Equation — Match Them Up One by One

1) There must always be the same number of atoms on both sides — they can't just disappear.
2) You balance the equation by putting numbers in front of the formulas where needed.

Take this equation for reacting sulfuric acid with sodium hydroxide:

$$H_2SO_4 + NaOH \rightarrow Na_2SO_4 + H_2O$$

The formulas are all correct but the numbers of some atoms don't match up on both sides.
You can't change formulas like H_2SO_4 to H_2SO_5. You can only put numbers in front of them:

Method: Balance Just ONE Type of Atom at a Time

The more you practise, the quicker you get, but all you do is this:

1) Find an element that doesn't balance and pencil in a number to try and sort it out.
2) See where it gets you. It may create another imbalance, but pencil in another number and see where that gets you.
3) Carry on chasing unbalanced elements and it'll sort itself out pretty quickly.

I'll show you. In the equation above you soon notice we're short of H atoms on the RHS (Right-Hand Side).
1) The only thing you can do about that is make it $2H_2O$ instead of just H_2O:

$$H_2SO_4 + NaOH \rightarrow Na_2SO_4 + 2H_2O$$

2) But that now causes too many H atoms and O atoms on the RHS, so to balance that up you could try putting $2NaOH$ on the LHS (Left-Hand Side):

$$H_2SO_4 + 2NaOH \rightarrow Na_2SO_4 + 2H_2O$$

3) And suddenly there it is! Everything balances. And you'll notice the Na just sorted itself out.

State Symbols Tell You What Physical State It's In

These are easy enough, so just make sure you know them — especially aq (aqueous).

(s) — Solid	(l) — Liquid	(g) — Gas	(aq) — Dissolved in water

E.g. $2Mg (s) + O_2 (g) \rightarrow 2MgO (s)$

All this balancing — it's enough to make you dizzy...

Write balanced symbol equations for these, and put the state symbols in too:
1) Iron(III) oxide + hydrogen → iron + water
2) Dilute hydrochloric acid + aluminium → aluminium chloride solution + hydrogen   Answers on page 71

Chemistry 2(i) — Bonding and Reactions

Relative Formula Mass

The biggest trouble with <u>relative atomic mass</u> and <u>relative formula mass</u> is that they <u>sound</u> so blood-curdling. Take a few deep breaths, and just enjoy, as the mists slowly clear...

Relative Atomic Mass, A_r — Easy Peasy

1) This is just a way of saying how <u>heavy</u> different atoms are <u>compared</u> with the mass of an atom of carbon-12. So carbon-12 has A_r of <u>exactly 12</u>.

2) It turns out that the <u>relative atomic mass</u> A_r is usually just the same as the <u>mass number</u> of the element.

3) In the periodic table, the elements all have <u>two</u> numbers. The smaller one is the atomic number (how many protons it has). But the <u>bigger one</u> is the <u>relative atomic mass</u>. Easy peasy, I'd say.

Relative Atomic Mass

$$^4_2\text{He} \qquad ^{12}_6\text{C} \qquad ^{35.5}_{17}\text{Cl}$$

> When an element has more than one stable isotope, the relative atomic mass is an average value of all the different isotopes (taking into account how much there is of each isotope).

Helium has $A_r = 4$. Carbon has $A_r = 12$. Chlorine has $A_r = 35.5$.

Relative Formula Mass, M_r — Also Easy Peasy

If you have a compound like $MgCl_2$ then it has a <u>relative formula mass</u>, M_r, which is just all the relative atomic masses <u>added together</u>.
For $MgCl_2$ it would be:

$$MgCl_2$$

$$24 \quad + \quad (35.5 \times 2) \quad = \quad 95$$

> So M_r for $MgCl_2$ is simply <u>95</u>.

You can easily get A_r for any element from the periodic table (see inside front cover), but in a lot of questions they give you them anyway. I tell you what, since it's nearly Christmas I'll run through another example for you:

> **Question:** Find the relative formula mass for calcium carbonate, $CaCO_3$, using the given data:
> A_r for Ca = 40 A_r for C = 12 A_r for O = 16

<u>ANSWER:</u>

$$CaCO_3$$

$$40 \quad + \quad 12 \quad + \quad (16 \times 3) = 100$$

> So the relative formula mass for $CaCO_3$ is <u>100</u>.

And that's all it is. A big fancy name like <u>relative formula mass</u> and all it means is "<u>add up all the mass numbers</u>". What a swizz, eh? You'd have thought it'd be something a bit juicier than that, wouldn't you. Still, that's life — it's all a big disappointment in the end. Sigh.

Numbers? — and you thought you were doing chemistry...

Learn the definitions of relative atomic mass and relative formula mass, then have a go at these:
1) Use the periodic table to find the relative atomic mass of these elements: Cu, K, Kr, Cl
2) Find the relative formula mass of: NaOH, Fe_2O_3, C_6H_{14}, $Mg(NO_3)_2$

Answers on page 71

Two Formula Mass Calculations

Although relative atomic mass and relative formula mass are <u>easy enough</u>, it can get just a tad <u>trickier</u> when you start getting into other calculations which use them. It depends on how good your maths is basically, because it's all to do with ratios and percentages.

Calculating % Mass of an Element in a Compound

This is actually dead easy — so long as you've learnt this formula:

$$\text{Percentage mass OF AN ELEMENT IN A COMPOUND} = \frac{A_r \times \text{No. of atoms (of that element)}}{M_r \quad \text{(of whole compound)}} \times 100$$

If you don't learn the formula then you'd better be pretty smart — or you'll struggle.

<u>EXAMPLE:</u> Find the percentage mass of sodium in sodium carbonate, Na_2CO_3.
<u>ANSWER:</u>
  A_r of sodium = 23, A_r of carbon = 12, A_r of oxygen = 16
  M_r of Na_2CO_3 = (2 × 23) + 12 + (3 × 16) = 106

Now use the formula: $\underline{\text{Percentage mass}} = \frac{A_r \times n}{M_r} \times 100 = \frac{23 \times 2}{106} \times 100 = 43.4\%$

And there you have it. Sodium makes up <u>43.4%</u> of the mass of sodium carbonate.

Finding the Empirical Formula (from Masses or Percentages)

This also sounds a lot worse than it really is. Try this for an easy peasy <u>stepwise method</u>:

1) <u>List all the elements</u> in the compound (there's usually only two or three!)
2) <u>Underneath them</u>, write their <u>experimental masses or percentages</u>.
3) <u>Divide</u> each mass or percentage <u>by the A_r</u> for that particular element.
4) Turn the numbers you get into <u>a nice simple ratio</u>
   by multiplying and/or dividing them by well-chosen numbers.
5) Get the ratio in its <u>simplest form</u>, and that tells you the empirical formula of the compound.

<u>Example:</u> Find the empirical formula of the iron oxide produced when 44.8 g of iron react with 19.2 g of oxygen. (A_r for iron = 56, A_r for oxygen =16)
<u>Method:</u>

1) List the two elements:	Fe	O
2) Write in the experimental masses:	44.8	19.2
3) Divide by the A_r for each element:	$^{44.8}/_{56} = 0.8$	$^{19.2}/_{16} = 1.2$
4) Multiply by 10...	8	12
...then divide by 4:	2	3

5) So the simplest formula is 2 atoms of Fe to 3 atoms of O, i.e. Fe_2O_3. And that's it done.

> You need to realise (for the exam) that this <u>empirical method</u> (i.e. based on <u>experiment</u>) is the <u>only way</u> of finding out the formula of a compound. Rust is iron oxide, sure, but is it FeO, or Fe_2O_3? Only an experiment to determine the empirical formula will tell you for certain.

With this empirical formula I can rule the world! — mwa ha ha ha...

Make sure you learn the formula and the five rules in the red box. Then try these: Answers on page 71
1) Find the percentage mass of oxygen in each of these: a) Fe_2O_3 b) H_2O c) $CaCO_3$ d) H_2SO_4.
2) Find the empirical formula of the compound formed from 2.4 g of carbon and 0.8 g of hydrogen.

Calculating Masses in Reactions

These can be kinda scary too, but chill out, little trembling one — just relax and enjoy.

The Three Important Steps — Not to Be Missed...

(Miss one out and it'll all go horribly wrong, believe me.)

1) Write out the balanced equation
2) Work out M_r — just for the two bits you want
3) Apply the rule: Divide to get one, then multiply to get all
   (But you have to apply this first to the substance they give information about, and then the other one!)

Don't worry — these steps should all make sense when you look at the example below.

Example: What mass of magnesium oxide is produced when 60 g of magnesium is burned in air?

Answer:

1) Write out the balanced equation:

$$2Mg + O_2 \rightarrow 2MgO$$

2) Work out the relative formula masses:

   (don't do the oxygen — we don't need it)

   $2 \times 24 \rightarrow 2 \times (24+16)$
   $48 \rightarrow 80$

3) Apply the rule: Divide to get one, then multiply to get all

   The two numbers, 48 and 80, tell us that   48 g of Mg react to give 80 g of MgO.
   Here's the tricky bit. You've now got to be able to write this down:

   > 48 g of Mg ......reacts to give.....80 g of MgO
   >
   > 1 g of Mg ......reacts to give.....
   >
   > 60 g of Mg .....reacts to give......

The big clue is that in the question they've said we want to burn "60 g of magnesium",
i.e. they've told us how much magnesium to have, and that's how you know to write down the left-hand side of it first, because:

   We'll first need to ÷ by 48 to get 1 g of Mg
   and then need to × by 60 to get 60 g of Mg.

Then you can work out the numbers on the other side (shown in orange below) by realising that you must divide both sides by 48 and then multiply both sides by 60. It's tricky.

÷48 { 48 g of Mg .... 80 g of MgO } ÷48
       1 g of Mg .... 1.67 g of MgO
×60 { 60 g of Mg .... 100 g of MgO } ×60

The mass of product is called the yield of a reaction. You should realise that in practice you never get 100% of the yield, so the amount of product will be slightly less than calculated (see p.52).

This finally tells us that 60 g of magnesium will produce 100 g of magnesium oxide.
If the question had said "Find how much magnesium gives 500 g of magnesium oxide", you'd fill in the MgO side first, because that's the one you'd have the information about. Got it? Good-O!

Reaction mass calculations — no worries, matey...

The only way to get good at these is to practise. So have a go at these: Answers on page 71
1) Find the mass of calcium which gives 30 g of calcium oxide (CaO) when burnt in air.
2) What mass of fluorine fully reacts with potassium to make 116 g of potassium fluoride (KF)?

The Mole

The mole is really confusing. I think it's the word that puts people off. It's very difficult to see the relevance of the word "mole" to different-sized piles of brightly coloured powders.

"THE MOLE" is Simply the Name Given to a Certain Number

Just like "a million" is this many: 1 000 000;   or "a billion" is this many: 1 000 000 000,
so "a mole" is this many: 602 300 000 000 000 000 000 000   or 6.023×10^{23}.

1) And that's all it is. Just a number. The burning question, of course, is why is it such a silly long one like that, and with a six at the front?
2) The answer is that when you get precisely that number of atoms of carbon-12 it weighs exactly 12 g. So, get that number of atoms or molecules, of any element or compound, and conveniently, they weigh exactly the same number of grams as the relative atomic mass, A_r (or M_r) of the element or compound.
This is arranged on purpose of course, to make things easier.

> **One mole of atoms or molecules of any substance will have a mass in grams equal to the relative formula mass (A_r or M_r) for that substance.**

Examples:
Iron has an A_r of 56.   So one mole of iron weighs exactly 56 g
Nitrogen gas, N_2, has an M_r of 28 (2×14).   So one mole of N_2 weighs exactly 28 g
Carbon dioxide, CO_2, has an M_r of 44.   So one mole of CO_2 weighs exactly 44 g

This means that 12 g of carbon, or 56 g of iron, or 28 g of N_2, or 44 g of CO_2, all contain the same number of particles, namely one mole or 6×10^{23} atoms or molecules.

Nice Easy Formula for Finding the Number of Moles in a Given Mass:

$$\text{NUMBER OF MOLES} = \frac{\text{Mass in g} \quad \text{(of element or compound)}}{M_r \quad \text{(of element or compound)}}$$

Example: How many moles are there in 42 g of carbon?
Answer: No. of moles   = Mass (g) / M_r = 42/12   = 3.5 moles   Easy Peasy

"Relative Formula Mass" is Also "Molar Mass"

1) We've been very happy using the relative formula mass, M_r all through the calculations.
2) In fact, that was already using the idea of Moles because M_r is actually the mass of one mole in g, or as we sometimes call it, the molar mass.
3) The volume of one mole of any gas at room temperature and pressure is 24 litres — the molar volume.

A "1M Solution" Contains "One Mole per Litre"

The 'moles per litre' of a solution is sometimes called its 'molarity'.

This is pretty easy. So a 2M solution of NaOH contains 2 moles of NaOH per litre of solution.
You need to know how many moles there'll be in a given volume:

$$\text{NUMBER OF MOLES} = \text{Volume in Litres} \times \text{Moles per Litre of solution}$$

Example: How many moles in 185 cm^3 of a 2M solution?   Ans: $0.185 \times 2 = 0.37$ moles

7 moles of moles ≈ 1 Earth... ...assuming vol. of 1 mole = ¼ litre, no gaps between moles, spherical Earth...

It's possible to do all the calculations on the previous pages without ever talking about moles.
You just concentrate on M_r and A_r instead — in fact M_r and A_r represent moles anyway. I think it's less confusing if moles aren't mentioned at all, but you need to know about them for the exam.

Atom Economy

It's important in industrial reactions that as much of the reactants as possible get turned into useful products. This depends on the atom economy and the percentage yield (see next page) of the reaction.

"Atom Economy" — % of Reactants Changed to Useful Products

1) A lot of reactions make more than one product. Some of them will be useful, but others will just be waste, e.g. when you make quicklime from limestone, you also get CO_2 as a waste product.

2) The atom economy of a reaction tells you how much of the mass of the reactants ends up as useful products. Learn the equation:

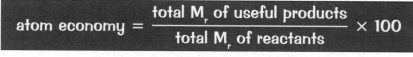

$$\text{atom economy} = \frac{\text{total } M_r \text{ of useful products}}{\text{total } M_r \text{ of reactants}} \times 100$$

Example: Hydrogen gas is made on a large scale by reacting natural gas (methane) with steam.

$$CH_4(g) + H_2O(g) \rightarrow CO(g) + 3H_2(g)$$

Calculate the atom economy of this reaction.

Answer:
1) Identify the useful product — that's the hydrogen gas.
2) Work out the M_r of reactants and the useful product:

CH_4	H_2O	$3H_2$
$12 + (4 \times 1)$	$(2 \times 1) + 16$	$3 \times (2 \times 1)$
34		6

3) Use the formula to calculate the atom economy: atom economy $= \dfrac{6}{34} \times 100 = \underline{17.6\%}$

So in this reaction, over 80% of the starting materials are wasted!

In industry, the waste CO is reacted with more steam to make CO_2 (and a bit more H_2). That brings the overall atom economy down to only 15% — but the final waste product is much less nasty that way.

High Atom Economy is Better for Profits and the Environment

1) Pretty obviously, if you're making lots of waste, that's a problem.

2) Reactions with low atom economy use up resources very quickly. At the same time, they make lots of waste materials that have to be disposed of somehow. That tends to make these reactions unsustainable — the raw materials will run out and the waste has to go somewhere.

3) For the same reasons, low atom economy reactions aren't usually profitable. Raw materials are expensive to buy, and waste products can be expensive to remove and dispose of responsibly.

4) The best way around the problem is to find a use for the waste products rather than just throwing them away. There's often more than one way to make the product you want, so the trick is to come up with a reaction that gives useful "by-products" rather than useless ones.

5) The reactions with the highest atom economy are the ones that only have one product — like the Haber process (see page 62). Those reactions have an atom economy of 100%.

So why do they make hydrogen in that nasty, inefficient way — well, at the moment it's the best of a bad bunch. The other ways to make hydrogen on an industrial scale (like the electrolysis of brine, see page 69) use up massive amounts of energy, and are too expensive to be worthwhile.

Atom economy — important but not the whole story...

You could get asked about any industrial reaction in the exam. Don't panic — whatever example they give you, the same stuff applies. In the real world, high atom economy isn't enough, though. You need to think about the percentage yield of the reaction (next page) and the energy cost as well.

Percentage Yield

Percentage yield tells you about the <u>overall success</u> of an experiment. It compares what you think you should get (<u>predicted yield</u>) with what you get in practice (<u>actual yield</u>).

Percentage Yield Compares Actual and Predicted Yield

The more reactants you start with, the higher the <u>actual yield</u> will be — that's pretty obvious. But the <u>percentage yield doesn't</u> depend on the amount of reactants you started with — it's a <u>percentage</u>.

1) The <u>predicted yield</u> of a reaction can be calculated from the <u>balanced reaction equation</u> (see page 49).

2) Percentage yield is given by the formula:

$$\text{percentage yield} = \frac{\text{actual yield (grams)}}{\text{predicted yield (grams)}} \times 100$$

3) Percentage yield is <u>always</u> somewhere between 0 and 100%.

4) A 100% percentage yield means that you got <u>all</u> the product you expected to get.

5) A 0% yield means that <u>no</u> reactants were converted into product, i.e. no product at all was <u>made</u>.

Yields Are Always Less Than 100%

In real life, you <u>never</u> get a 100% percentage yield. Some product or reactant <u>always</u> gets lost along the way — and that goes for big <u>industrial processes</u> as well as school lab experiments. How this happens depends on <u>what sort of reaction</u> it is and what <u>apparatus</u> is being used.

Lots of things can go wrong, but the four you need to <u>know about</u> are:

1) The Reaction is Reversible

In <u>reversible reactions</u> (like the Haber process, see page 62), not all the reactants change into product.

Instead, you get <u>reactants</u> and <u>products</u> in <u>equilibrium</u>. Increasing the temperature moves the <u>equilibrium position</u> (see page 61), so heating the reaction to speed it up might mean a <u>lower yield</u>.

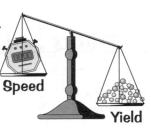

2) Filtration

When you <u>filter a liquid</u> to remove <u>solid particles</u>, you nearly always lose a bit of liquid or a bit of solid.

1) If you want to <u>keep the liquid</u>, you lose the bit that remains with the solid and filter paper (as they always stay a bit wet).

2) If you want to <u>keep the solid</u>, some of it usually gets left behind when you scrape it off the filter paper — even if you're really careful.

3) Transferring Liquids

You always lose a bit of liquid when you <u>transfer</u> it from one container to another — even if you manage not to spill it.

Some of it always gets left behind on the <u>inside surface</u> of the old container. Think about it — it's always wet when you finish.

4) Unexpected Reactions

Things don't always go exactly to plan.

Sometimes you get unexpected reactions happening, so the yield of the <u>intended product</u> goes down. These can be caused by <u>impurities</u> in the reactants, but sometimes just changing the <u>reaction conditions</u> affects what products you make.

You can't always get what you want...

A high percentage yield means there's <u>not much waste</u> — which is good for <u>preserving resources</u>, and keeping production <u>costs down</u>. If a reaction's going to be worth doing commercially, it generally has to have a high percentage yield or recyclable reactants, e.g. the Haber process.

Revision Summary for Chemistry 2(i)

Now, my spies tell me that some naughty people skip these pages without so much as reading through the list of questions. Well, you shouldn't, because what's the point in reading that great big section if you're not going to check if you really know it or not? Look, just read the first ten questions, and I guarantee there'll be an answer you'll have to look up. And when it comes up in the exam, you'll be so glad you did. Plus, if you don't do as you're told my spies will tell me, and then you won't get any toys. Or something.

1) Sketch the nuclear model of an atom.
    Give five details about the nucleus and five details about the electrons.
2) Draw a table showing the relative masses and charges of the three types of particle in an atom.
3) What do the mass number and atomic number represent?
4) Explain the difference between an element and a compound. Give an example of each.
5) Define the term isotope.
6) What feature of atoms determines the order of the modern periodic table?
7) What are groups in the periodic table? Explain their significance in terms of electrons.
8) Describe how you would work out the electron configuration of an atom given its atomic number.
    Find the electron configuration of potassium (using the periodic table at the front of the book).
9) Describe the process of ionic bonding.
10) List the main properties of ionic compounds.
11) What is covalent bonding?
12) What are the two types of covalent substance? Give three examples of each.
13) Industrial diamonds are used in drill tips and precision cutting tools. What property of diamond makes it suitable for this use? Explain how the bonding in diamond causes its physical properties.
14) List three properties of metals and explain how metallic bonding causes these properties.
15)* a) Identify the type of bonding in each of the substances in the table:

Substance	Melting point (°C)	Electrical conductivity	Hardness [scale of 0 – 10 (10 being diamond)]
A	3410	Very high	7.5
B	2072	Zero	9
C	605	Zero in solid form High when molten	Low

b) Suggest which substance from the table would be most suitable for the following applications:
    i) a light-bulb filament
    ii) abrasive paper
    iii) a rechargeable battery

16) Give an example of a "smart" material and describe how it behaves.
17) What are nanoparticles? Describe two different applications of nanoparticles.
18)* Write balanced symbol equations for the following reactions:
    a) calcium carbonate + hydrochloric acid → calcium chloride + water + carbon dioxide
    b) sulfuric acid + potassium hydroxide → potassium sulfate + water
19) Define relative atomic mass and relative formula mass.
20)* Find A_r or M_r for these (use the periodic table at the front of the book):
    a) Ca   b) Ag   c) CO_2 d) $MgCO_3$ e) Na_2CO_3 f) ZnO   g) KOH   h) NH_3
21)* a) Calculate the percentage mass of carbon in: i) $CaCO_3$ ii) CO_2 iii) CH_4
    b) Calculate the percentage mass of metal in: i) Na_2O ii) Fe_2O_3 iii) Al_2O_3
22)* What is an empirical formula? Find the empirical formula of the compound formed when 21.9 g of magnesium, 29.3 g of sulfur and 58.4 g of oxygen react.
23)* What mass of sodium is needed to produce 108.2 g of sodium oxide (Na_2O)?
24) What is a mole? Why is it that precise number?
25)* How many moles of HCl are there in 230 cm^3 of 2M hydrochloric acid?
26) Explain why it is important, both economically and environmentally, to use industrial reactions with a high atom economy.
27) Describe four factors that can reduce the percentage yield of a reaction.

* Answers on page 71

Rates of Reaction

Reactions Can Go at All Sorts of Different Rates

1) One of the <u>slowest</u> is the <u>rusting</u> of iron (it's not slow enough though — what about my little MGB).

2) A <u>moderate speed</u> reaction is a <u>metal</u> (like magnesium) reacting with <u>acid</u> to produce a gentle stream of <u>bubbles</u>.

3) A <u>really fast</u> reaction is an <u>explosion</u>, where it's all over in a <u>fraction</u> of a second.

The Rate of a Reaction Depends on Four Things:

1) <u>Temperature</u>
2) <u>Concentration</u> — (or <u>pressure</u> for gases)
3) <u>Catalyst</u>
4) <u>Size of particles</u> — (or <u>surface area</u>)

LEARN THEM!

Typical Graphs for Rate of Reaction

The plot below shows how the speed of a particular reaction varies under <u>different conditions</u>. The quickest reaction is shown by the line that becomes <u>flat</u> in the <u>least</u> time. The line that flattens out first must have the <u>steepest slope</u> compared to all the others, making it possible to spot the slowest and fastest reactions.

1) <u>Graph 1</u> represents the original <u>fairly slow</u> reaction. The graph is not too steep.

2) <u>Graphs 2 and 3</u> represent the reaction taking place <u>quicker</u> but with the <u>same initial amounts</u>. The slope of the graphs gets steeper.

3) The <u>increased rate</u> could be due to <u>any</u> of these:

   a) increase in <u>temperature</u>
   b) increase in <u>concentration</u> (or pressure)
   c) <u>catalyst</u> added
   d) solid reactant crushed up into <u>smaller bits</u>.

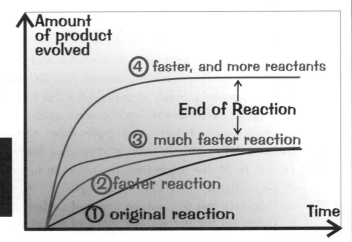

4) <u>Graph 4</u> produces <u>more product</u> as well as going <u>faster</u>. This can <u>only</u> happen if <u>more reactant(s)</u> are added at the start. <u>Graphs 1, 2 and 3</u> all converge at the same level, showing that they all produce the same amount of product, although they take <u>different</u> times to get there.

How to get a fast, furious reaction — crack a wee joke...

<u>Industrial</u> reactions generally use a <u>catalyst</u> and are done at <u>high temperature and pressure</u>. Time is money, so the faster an industrial reaction goes the better... but only <u>up to a point</u>. Chemical plants are quite expensive to rebuild if they get blown into lots and lots of teeny tiny pieces.

Measuring Rates of Reaction

Three Ways to Measure the Speed of a Reaction

The <u>speed of a reaction</u> can be observed <u>either</u> by how quickly the reactants are used up or how quickly the products are formed. It's usually a lot easier to measure <u>products forming</u>.

The rate of reaction can be calculated using the following equation:

$$\text{Rate of Reaction} = \frac{\text{Amount of reactant used or amount of product formed}}{\text{Time}}$$

There are different ways that the speed of a reaction can be <u>measured</u>. Learn these three:

1) Precipitation

1) This is when the product of the reaction is a <u>precipitate</u> which <u>clouds</u> the solution.
2) Observe a <u>marker</u> through the solution and measure how long it takes for it to <u>disappear</u>.
3) The <u>quicker</u> the marker disappears, the <u>quicker</u> the reaction.
4) This only works for reactions where the initial solution is rather <u>see-through</u>.
5) The result is very <u>subjective</u> — <u>different people</u> might not agree over the <u>exact</u> point when the mark 'disappears'.

2) Change in Mass (Usually Gas Given Off)

1) Measuring the speed of a reaction that <u>produces a gas</u> can be carried out on a <u>mass balance</u>.
2) As the gas is released the mass <u>disappearing</u> is easily measured on the balance.
3) The <u>quicker</u> the reading on the balance <u>drops</u>, the <u>faster</u> the reaction.
4) <u>Rate of reaction graphs</u> are particularly easy to plot using the results from this method.
5) This is the <u>most accurate</u> of the three methods described on this page because the mass balance is very accurate. But it has the <u>disadvantage</u> of releasing the gas straight into the room.

3) The Volume of Gas Given Off

1) This involves the use of a <u>gas syringe</u> to measure the <u>volume</u> of gas given off.
2) The <u>more</u> gas given off during a given <u>time interval</u>, the <u>faster</u> the reaction.
3) A graph of <u>gas volume</u> against <u>time elapsed</u> could be plotted to give a rate of reaction graph.
4) Gas syringes usually give volumes accurate to the <u>nearest millilitre</u>, so they're quite accurate. You have to be quite careful though — if the reaction is too <u>vigorous</u>, you can easily blow the plunger out of the end of the syringe!

OK have you got your stopwatch ready *BANG!* — oh...

Each method has its <u>pros and cons</u>. The mass balance method is only accurate as long as the flask isn't too hot, otherwise you lose mass by evaporation as well as by the reaction. The first method isn't very accurate, but if you're not producing a gas you can't use either of the other two. Ah well.

Rate of Reaction Experiments

Remember: Any reaction can be used to investigate any of the four factors that affect the rate. These pages illustrate four important reactions, but only one factor has been considered for each. But we could just as easily use, say, the marble chips/acid reaction to test the effect of temperature instead.

1) Reaction of Hydrochloric Acid and Marble Chips

This experiment is often used to demonstrate the effect of breaking the solid up into small bits.

1) Measure the volume of gas evolved with a gas syringe and take readings at regular intervals.

2) Make a table of readings and plot them as a graph. You choose regular time intervals, so time is the independent variable (x) and volume is the dependent variable (y).

3) Repeat the experiment with exactly the same volume of acid, and exactly the same mass of marble chips, but with the marble more crunched up.

4) Then repeat with the same mass of powdered chalk instead of marble chips.

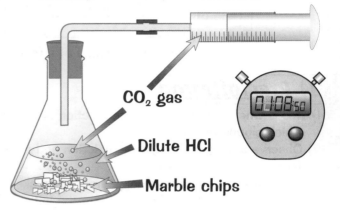

CO₂ gas

Dilute HCl

Marble chips

This graph shows the effect of using finer particles of solid

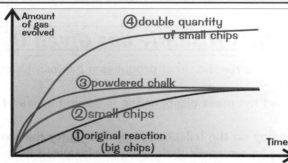

Amount of gas evolved

④ double quantity of small chips
③ powdered chalk
② small chips
① original reaction (big chips)

Time

1) An increase in surface area causes more collisions, so the rate of reaction is faster.

2) Line 4 shows the reaction if a greater mass of small marble chips is added.

3) The extra surface area gives a quicker reaction and there is also more gas evolved overall.

2) Reaction of Magnesium Metal with Dilute HCl

1) This reaction is good for measuring the effects of increased concentration (as is the marble/acid reaction).

2) This reaction gives off hydrogen gas, which we can measure with a mass balance, as shown.

3) In this experiment, time is again the independent variable and mass loss is the dependent variable. (The other method is to use a gas syringe, as above.)

This graph shows the effect of using more concentrated acid solutions

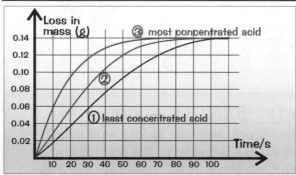

Loss in mass (g)

0.14
0.12
0.10
0.08
0.06
0.04
0.02

③ most concentrated acid
②
① least concentrated acid

Time/s

10 20 30 40 50 60 70 80 90 100

1) Take readings of mass at regular time intervals.

2) Put the results in a table and work out the loss in mass for each reading. Plot a graph.

3) Repeat with more concentrated acid solutions, but always with the same amount of magnesium.

4) The volume of acid must always be kept the same too — only the concentration is increased.

5) The three graphs show the same old pattern — a higher concentration giving a steeper graph, with the reaction finishing much quicker.

Chemistry 2(ii) — Rates of Reaction

Rate of Reaction Experiments

3) Sodium Thiosulfate and HCl Produce a Cloudy Precipitate

1) These two chemicals are both <u>clear solutions</u>.
2) They react together to form a <u>yellow precipitate</u> of <u>sulfur</u>.
3) The experiment involves watching a black mark <u>disappear</u> through the <u>cloudy sulfur</u> and <u>timing</u> how long it takes to go.

4) The reaction can be <u>repeated</u> for solutions at different <u>temperatures</u>. In practice, that's quite hard to do accurately and safely (it's not a good idea to heat an acid directly). The best way to do it is to use a <u>water bath</u> to heat both solutions to the right temperature <u>before you mix them</u>.
5) The <u>depth</u> of liquid must be kept the <u>same</u> each time, of course.
6) The results will of course show that the <u>higher</u> the temperature the <u>quicker</u> the reaction and therefore the <u>less time</u> it takes for the mark to <u>disappear</u>. These are typical results:

Temperature (°C)	20	25	30	35	40
Time taken for mark to disappear (s)	193	151	112	87	52

independent variable

dependent variable

This reaction can <u>also</u> be used to test the effects of <u>concentration</u>.
One sad thing about this reaction is it <u>doesn't</u> give a set of graphs. Well I think it's sad. All you get is a set of <u>readings</u> of how long it took till the mark disappeared for each temperature. Boring.

4) The Decomposition of Hydrogen Peroxide

This is a <u>good</u> reaction for showing the effect of different <u>catalysts</u>.
The decomposition of hydrogen peroxide is:

$$2H_2O_{2(aq)} \rightleftharpoons 2H_2O_{(l)} + O_{2(g)}$$

1) This is normally quite <u>slow</u> but a sprinkle of <u>manganese(IV) oxide catalyst</u> speeds it up no end. Other catalysts which work are found in:
   a) <u>potato peel</u> and b) <u>blood</u>.
2) <u>Oxygen gas</u> is given off, which provides an <u>ideal way</u> to measure the rate of reaction using the good ol' <u>gas syringe</u> method.

O_2 gas

Hydrogen peroxide

Catalyst

1) Same old graphs of course.
2) <u>Better</u> catalysts give a <u>quicker reaction</u>, which is shown by a <u>steeper graph</u> which levels off quickly.
3) This reaction can also be used to measure the effects of <u>temperature</u>, or of <u>concentration</u> of the H_2O_2 solution. The graphs will look just the same.

BLOOD is a catalyst? — eeurgh...

You don't need to know all the details of these specific reactions — it's the experimental methods you need to learn. If you understand how all this works, you should be able to apply it to any reaction.

Collision Theory

Reaction rates are explained perfectly by collision theory. It's really simple. It just says that the rate of a reaction simply depends on how often and how hard the reacting particles collide with each other. The basic idea is that particles have to collide in order to react, and they have to collide hard enough (with enough energy).

More Collisions Increases the Rate of Reaction

All four methods of increasing the rate of reactions can be explained in terms of increasing the number of successful collisions between the reacting particles:

1) HIGHER TEMPERATURE increases collisions

When the temperature is increased the particles all move quicker. If they're moving quicker, they're going to have more collisions.

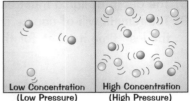

2) HIGHER CONCENTRATION (or PRESSURE) increases collisions

If a solution is made more concentrated it means there are more particles of reactant knocking about between the water molecules which makes collisions between the important particles more likely. In a gas, increasing the pressure means the particles are more squashed up together so there are going to be more collisions.

Low Concentration (Low Pressure)   High Concentration (High Pressure)

3) LARGER SURFACE AREA increases collisions

If one of the reactants is a solid then breaking it up into smaller pieces will increase its surface area. This means the particles around it in the solution will have more area to work on, so there'll be more useful collisions.

4) CATALYSTS increase the number of SUCCESSFUL collisions

A solid catalyst works by giving the reacting particles a surface to stick to. They increase the number of SUCCESSFUL collisions by lowering the activation energy (see next page).

Surface of catalyst

Faster Collisions Increase the Rate of Reaction

Higher temperature also increases the energy of the collisions, because it makes all the particles move faster.

Faster collisions are ONLY caused by increasing the temperature

Reactions only happen if the particles collide with enough energy.

At a higher temperature there will be more particles colliding with enough energy to make the reaction happen.

This initial energy is known as the activation energy, and it's needed to break the initial bonds.

Cool Atoms   Hot Atoms

Collision theory — the lamppost ran into me...

Once you've learnt everything off this page, the rates of reaction stuff should start making a lot more sense to you. The concept's fairly simple — the more often particles bump into each other, and the harder they hit when they do, the faster the reaction happens.

Catalysts

Many reactions can be speeded up by adding a catalyst.

A catalyst is a substance which changes the speed of a reaction, without being changed or used up in the reaction.

1) Catalysts Lower the Activation Energy

1) The activation energy is the minimum amount of energy needed for a reaction to happen.

2) It's a bit like having to climb up one side of a hill before you can ski/snowboard/sledge/fall down the other side.

3) Catalysts lower the activation energy of reactions, making it easier for them to happen.

4) This means a lower temperature can be used.

Energy

Without Catalyst

Activation Energy Without Catalyst

With Catalyst

Reactants

Activation Energy When Catalyst Present

Products

ΔH

Progress of Reaction

2) Solid Catalysts Work Best When They Have a Big Surface Area

1) Catalysts are usually used as a powder or pellets or a fine gauze.

2) This gives them a very large surface area to enable the reacting particles to meet up and do the business.

Catalyst Powder   Catalyst Pellets   Catalyst Gauzes

3) Transition metals are common catalysts in many industrial reactions, e.g. nickel is used for cracking hydrocarbons and iron catalyses the Haber process (see p.62).

3) Catalysts Help Reduce Costs in Industrial Reactions

1) Catalysts are very important for commercial reasons — most industrial reactions use them.

2) Catalysts increase the rate of the reaction, which saves a lot of money simply because the plant doesn't need to operate for as long to produce the same amount of stuff.

3) Alternatively, a catalyst will allow the reaction to work at a much lower temperature. That reduces the energy used up in the reaction (the energy cost), which is good for sustainable development and can save a lot of money too.

4) There are disadvantages to using catalysts, though.

5) They can be very expensive to buy, and often need to be removed from the product and cleaned. They never get used up in the reaction though, so once you've got them you can use them over and over again.

6) Different reactions use different catalysts, so if you make more than one product at your plant, you'll probably need to buy different catalysts for them.

7) Catalysts can be 'poisoned' by impurities, so they stop working, e.g. sulfur can poison the iron catalyst used in the Haber process. That means you have to keep your reaction mixture very clean.

Catalysts are like great jokes — they can be used over and over...

And they're not only used in industry... every useful chemical reaction in the human body is catalysed by a biological catalyst (an enzyme). If the reactions in the body were just left to their own devices, they'd take so long to happen, we couldn't exist. Quite handy then, these catalysts.

Energy Transfer in Reactions

Whenever chemical reactions occur <u>energy</u> is usually <u>transferred</u> to or from the <u>surroundings</u>.

In an Exothermic Reaction, Heat is Given Out

An <u>EXOTHERMIC reaction</u> is one which <u>gives out energy</u> to the surroundings, usually in the form of <u>heat</u> and usually shown by a <u>rise in temperature.</u>

1) Burning Fuels

The best example of an <u>exothermic</u> reaction is <u>burning fuels</u> — also called <u>COMBUSTION</u>. This gives out a lot of heat — it's very exothermic.

2) Neutralisation Reactions

<u>Neutralisation reactions</u> (acid + alkali) are also exothermic — see page 64.

3) Oxidation Reactions

Many oxidation reactions are exothermic...

Addition of sodium to water <u>produces heat</u>, so it must be <u>exothermic</u>. The sodium emits <u>heat</u> and moves about on the surface of the water as it is oxidised.

ACID

<u>Don't</u> do it like this!!

ALKALI

In an Endothermic Reaction, Heat is Taken In

An <u>ENDOTHERMIC reaction</u> is one which <u>takes in energy</u> from the surroundings, usually in the form of <u>heat</u> and usually shown by a <u>fall in temperature.</u>

Endothermic reactions are much <u>less common</u>. <u>Thermal decompositions</u> are a good example:

<u>THERMAL DECOMPOSITION OF CALCIUM CARBONATE</u>:

Heat must be supplied to make the compound <u>decompose</u> to make quicklime.

$$CaCO_3 \rightarrow CaO + CO_2$$

<u>A lot of heat energy</u> is needed to make this happen. In fact the calcium carbonate has to be <u>heated in a kiln</u> and kept at about <u>800 °C</u>. It takes almost <u>18 000 kJ</u> of heat to make <u>10 kg</u> of calcium carbonate decompose. That's pretty endothermic I'd say.

The Thermal Decomposition of Hydrated Copper Sulfate

<u>Copper(II) sulfate</u> crystals can be used as a <u>test</u> for <u>water</u>.

1) If you <u>heat</u> <u>blue hydrated</u> copper(II) sulfate crystals it drives the water off and leaves <u>white anhydrous</u> copper(II) sulfate powder. This is endothermic.

Water vapour

2) If you then <u>add</u> a couple of drops of <u>water</u> to the <u>white powder</u> you get the <u>blue crystals</u> back again. This is exothermic.

This is a <u>reversible reaction</u> (see p.61). In reversible reactions, if the reaction is <u>endothermic</u> in <u>one direction</u>, it will be <u>exothermic</u> in the <u>other direction</u>. The energy absorbed by the endothermic reaction is <u>equal</u> to the energy released during the exothermic reaction.

Right, so burning gives out heat — really...

This whole energy transfer thing is a fairly simple idea — don't be put off by the long words. Remember, "<u>exo-</u>" = <u>exit</u>, "<u>-thermic</u>" = <u>heat</u>, so an exothermic reaction is one that <u>gives out</u> heat. And "<u>endo-</u>" = erm... the other one. Okay, so there's no easy way to remember that one. Tough.

Reversible Reactions

A <u>reversible reaction</u> is one where the <u>products</u> of the reaction can react with each other and <u>convert back</u> to the original reactants. In other words, <u>it can go both ways</u>.

> A <u>reversible reaction</u> is one where the <u>products</u> of the
> reaction can <u>themselves react</u> to produce the <u>original reactants</u>
> $$A + B \rightleftharpoons C + D$$

Reversible Reactions Will Reach Dynamic Equilibrium

1) If a reversible reaction takes place in a <u>closed system</u> then a state of <u>equilibrium</u> will always be reached.

2) <u>Equilibrium</u> means that the <u>relative (%) quantities</u> of reactants and products will reach a certain <u>balance</u> and stay there. (A '<u>closed system</u>' just means that none of the reactants or products can <u>escape</u>.)

3) It is in fact a <u>DYNAMIC EQUILIBRIUM</u>, which means that the reactions are still taking place in <u>both directions</u>, but the <u>overall effect is nil</u> because the forward and reverse reactions <u>cancel</u> each other out. The reactions are taking place at <u>exactly the same rate</u> in both directions.

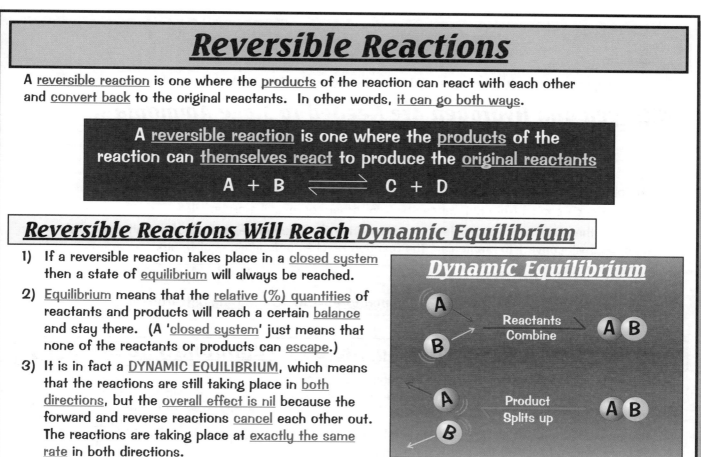

Dynamic Equilibrium

Reactants Combine

Product Splits up

Changing Temperature and Pressure to Get More Product

1) In a reversible reaction the '<u>position of equilibrium</u>' (the relative amounts of reactants and products) depends <u>very strongly</u> on the <u>temperature</u> and <u>pressure</u> surrounding the reaction.

2) If you <u>deliberately alter</u> the temperature and pressure you can <u>move</u> the "position of equilibrium" to give <u>more product</u> and <u>less</u> reactants.

Temperature

All reactions are <u>exothermic</u> in one direction and <u>endothermic</u> in the other.
 If you <u>raise</u> the <u>temperature</u>, the <u>endothermic</u> reaction will increase to <u>use up</u> the extra heat.
 If you <u>reduce</u> the <u>temperature</u> the <u>exothermic</u> reaction will increase to <u>give out</u> more heat.

Pressure

Many reactions have a <u>greater volume</u> on one side, either of <u>products</u> or <u>reactants</u> (greater volume means there are more molecules and less volume means there are fewer molecules).
 If you <u>raise</u> the <u>pressure</u> it will encourage the reaction which produces <u>less volume</u>.
 If you <u>lower</u> the <u>pressure</u> it will encourage the reaction which produces <u>more volume</u>.

<u>Adding a CATALYST doesn't change the equilibrium position:</u>

1) Catalysts speed up <u>both</u> the <u>forward</u> and <u>backward</u> reactions by the <u>same amount</u>.

2) So, adding a catalyst means the reaction reaches equilibrium <u>quicker</u>, but you end up with the <u>same amount</u> of product as you would without the catalyst.

Remember — catalysts DON'T affect the equilibrium position...

Changing the temperature <u>always</u> changes the equilibrium position, but that's not true of pressure. If your reaction has the same number of molecules on each side of the equation, changing the pressure won't make any difference at all to the equilibrium position (it still affects the <u>rate</u> of reaction though).

The Haber Process

This is an important industrial process. It produces ammonia (NH_3), which is used to make fertilisers.

Nitrogen and Hydrogen Are Needed to Make Ammonia

$$N_{2\,(g)} + 3H_{2\,(g)} \rightleftharpoons 2NH_{3\,(g)} \quad (+ \text{ heat})$$

1) The nitrogen is obtained easily from the air, which is 78% nitrogen (and 21% oxygen).

2) The hydrogen comes from natural gas or from other sources like crude oil.

3) Because the reaction is reversible — it occurs in both directions — not all of the nitrogen and hydrogen will convert to ammonia. The reaction reaches a dynamic equilibrium.

Industrial conditions:
Pressure: 200 atmospheres;   Temperature: 450 °C;   Catalyst: Iron

The Reaction is Reversible, So There's a Compromise to be Made:

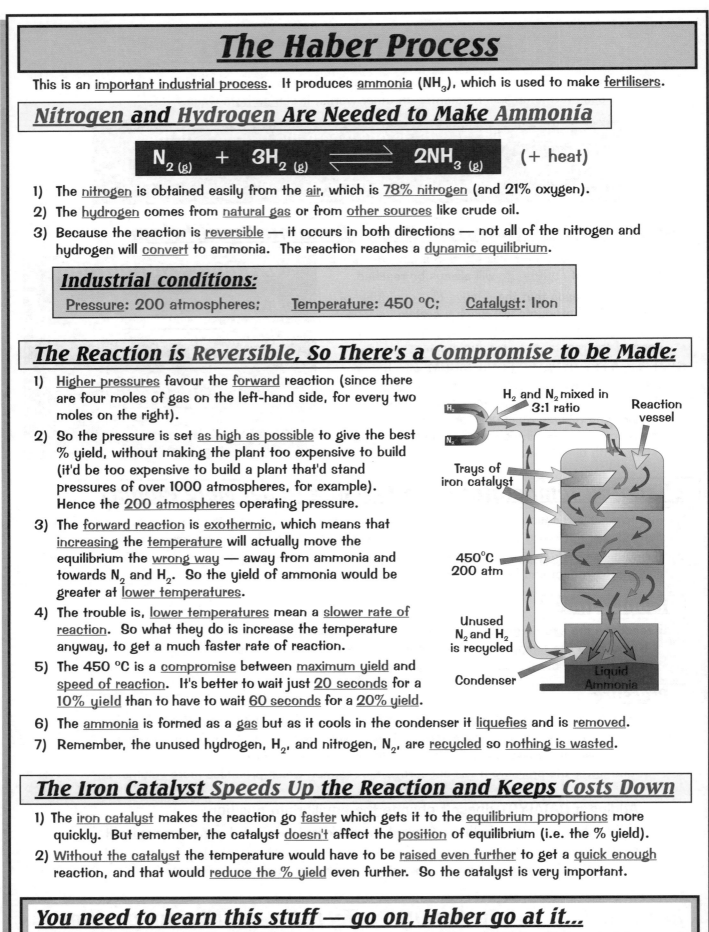

1) Higher pressures favour the forward reaction (since there are four moles of gas on the left-hand side, for every two moles on the right).

2) So the pressure is set as high as possible to give the best % yield, without making the plant too expensive to build (it'd be too expensive to build a plant that'd stand pressures of over 1000 atmospheres, for example). Hence the 200 atmospheres operating pressure.

3) The forward reaction is exothermic, which means that increasing the temperature will actually move the equilibrium the wrong way — away from ammonia and towards N_2 and H_2. So the yield of ammonia would be greater at lower temperatures.

4) The trouble is, lower temperatures mean a slower rate of reaction. So what they do is increase the temperature anyway, to get a much faster rate of reaction.

5) The 450 °C is a compromise between maximum yield and speed of reaction. It's better to wait just 20 seconds for a 10% yield than to have to wait 60 seconds for a 20% yield.

6) The ammonia is formed as a gas but as it cools in the condenser it liquefies and is removed.

7) Remember, the unused hydrogen, H_2, and nitrogen, N_2, are recycled so nothing is wasted.

Figure labels: H_2 and N_2 mixed in 3:1 ratio; Reaction vessel; Trays of iron catalyst; 450°C 200 atm; Unused N_2 and H_2 is recycled; Condenser; Liquid Ammonia

The Iron Catalyst Speeds Up the Reaction and Keeps Costs Down

1) The iron catalyst makes the reaction go faster which gets it to the equilibrium proportions more quickly. But remember, the catalyst doesn't affect the position of equilibrium (i.e. the % yield).

2) Without the catalyst the temperature would have to be raised even further to get a quick enough reaction, and that would reduce the % yield even further. So the catalyst is very important.

You need to learn this stuff — go on, Haber go at it...

The trickiest bit is remembering that the temperature is raised not for a better equilibrium, but for speed. It doesn't matter that the percentage yield is low, because the hydrogen and nitrogen are recycled. Cover the page and scribble down as much as you can remember, then check, and try again.

Revision Summary for Chemistry 2(ii)

Well, I don't think that was too bad, was it. Four things affect the rate of reactions, there are loads of ways to measure reaction rates and it's all explained by collision theory. Reactions can be endothermic or exothermic, and quite a few of them are reversible. Easy. Ahem.

Well here are some more of those nice questions that you enjoy so much. If there are any you can't answer, go back to the appropriate page, do a bit more learning, then try again.

1) What are the four factors that affect the rate of a reaction?

2) Describe three different ways of measuring the rate of a reaction. List one advantage and one disadvantage of each method.

3) A student carries out an experiment to measure the effect of surface area on the reaction between marble and hydrochloric acid. He measures the amount of gas given off at regular intervals.
   a) What factors must he keep constant for it to be a fair test?
   b)* He uses four samples for his experiment:
      Sample A – 10 g of powdered marble
      Sample B – 10 g of small marble chips
      Sample C – 10 g of large marble chips
      Sample D – 5 g of powdered marble
   Sketch a typical set of graphs for this experiment.

4) Explain how each of the four factors which affect reaction rates increases the number of successful collisions between particles.

5) What is the other aspect of collision theory which determines the rate of reaction?

6) Which is the only physical factor which affects this other aspect of collision theory?

7) What is the definition of a catalyst? What does a catalyst do to the activation energy of a reaction?

8) Discuss the advantages and disadvantages of using catalysts in industrial processes.

9) What is an exothermic reaction? Give three examples.

10) The reaction to split ammonium chloride into ammonia and hydrogen chloride is endothermic. What can you say for certain about the reverse reaction?

11) What is a reversible reaction? Explain what is meant by a dynamic equilibrium.

12) How does changing the temperature and pressure of a reversible reaction alter the equilibrium position?

13) How does this influence the choice of pressure for the Haber process?

14) What determines the choice of operating temperature for the Haber process?

15) What effect does the iron catalyst have on the reaction between nitrogen and hydrogen?

* Answers on page 71

Acids and Alkalis

The pH Scale and Universal Indicator

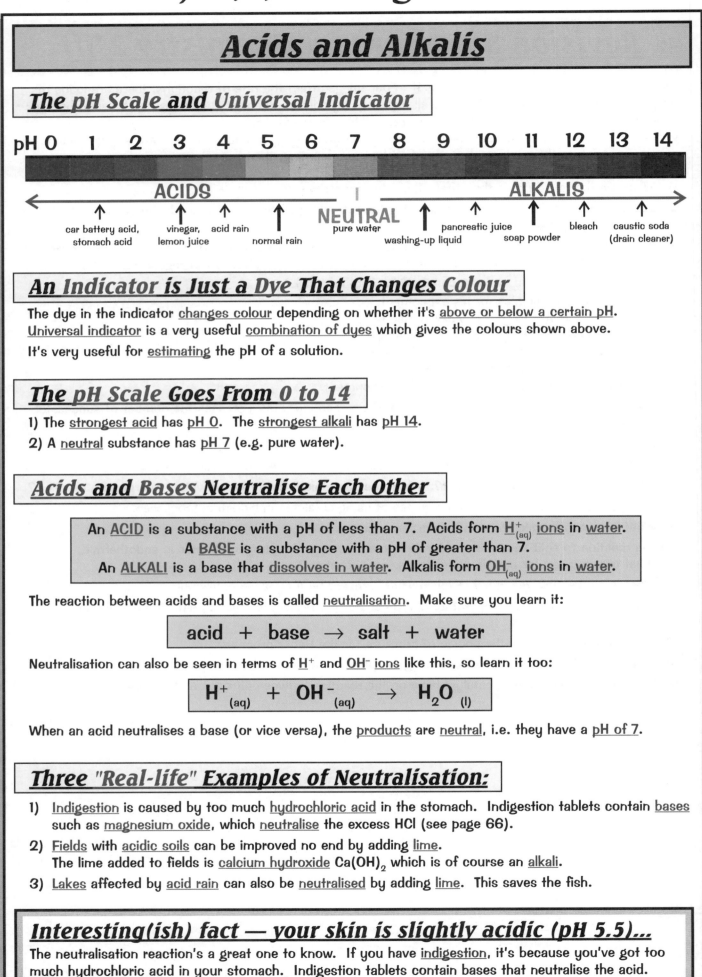

pH 0 1 2 3 4 5 6 7 8 9 10 11 12 13 14

ACIDS | NEUTRAL | ALKALIS

car battery acid, stomach acid — vinegar, lemon juice — acid rain — normal rain — pure water — washing-up liquid — pancreatic juice — soap powder — bleach — caustic soda (drain cleaner)

An Indicator is Just a Dye That Changes Colour

The dye in the indicator <u>changes colour</u> depending on whether it's <u>above or below a certain pH</u>. <u>Universal indicator</u> is a very useful <u>combination of dyes</u> which gives the colours shown above. It's very useful for <u>estimating</u> the pH of a solution.

The pH Scale Goes From 0 to 14

1) The <u>strongest acid</u> has <u>pH 0</u>. The <u>strongest alkali</u> has <u>pH 14</u>.
2) A <u>neutral</u> substance has <u>pH 7</u> (e.g. pure water).

Acids and Bases Neutralise Each Other

An <u>ACID</u> is a substance with a pH of less than 7. Acids form $H^+_{(aq)}$ <u>ions</u> in <u>water</u>.
A <u>BASE</u> is a substance with a pH of greater than 7.
An <u>ALKALI</u> is a base that <u>dissolves in water</u>. Alkalis form $OH^-_{(aq)}$ <u>ions</u> in <u>water</u>.

The reaction between acids and bases is called <u>neutralisation</u>. Make sure you learn it:

$$acid + base \rightarrow salt + water$$

Neutralisation can also be seen in terms of <u>H$^+$</u> and <u>OH$^-$</u> <u>ions</u> like this, so learn it too:

$$H^+_{(aq)} + OH^-_{(aq)} \rightarrow H_2O_{(l)}$$

When an acid neutralises a base (or vice versa), the <u>products</u> are <u>neutral</u>, i.e. they have a <u>pH of 7</u>.

Three "Real-life" Examples of Neutralisation:

1) <u>Indigestion</u> is caused by too much <u>hydrochloric acid</u> in the stomach. Indigestion tablets contain <u>bases</u> such as <u>magnesium oxide</u>, which <u>neutralise</u> the excess HCl (see page 66).
2) <u>Fields</u> with <u>acidic soils</u> can be improved no end by adding <u>lime</u>.
   The lime added to fields is <u>calcium hydroxide</u> Ca(OH)$_2$ which is of course an <u>alkali</u>.
3) <u>Lakes</u> affected by <u>acid rain</u> can also be <u>neutralised</u> by adding <u>lime</u>. This saves the fish.

Interesting(ish) fact — your skin is slightly acidic (pH 5.5)...

The neutralisation reaction's a great one to know. If you have <u>indigestion</u>, it's because you've got too much hydrochloric acid in your stomach. Indigestion tablets contain bases that neutralise the acid.

Acids Reacting with Metals

Acid + Metal → Salt + Hydrogen

That's written big 'cos it's kinda worth remembering. Here's the <u>typical experiment</u>:

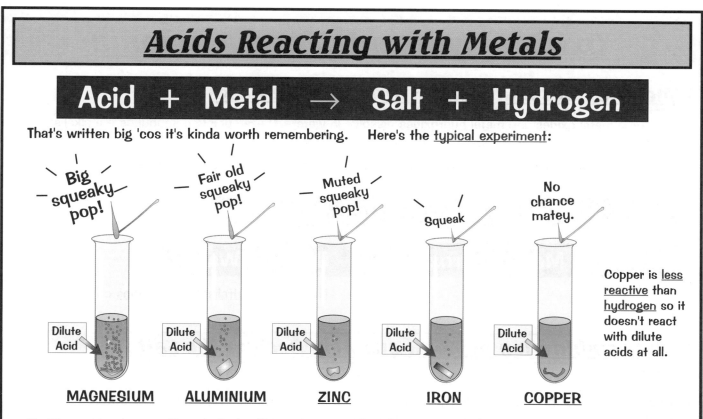

- Big squeaky pop!
- Fair old squeaky pop!
- Muted squeaky pop!
- Squeak
- No chance matey.

Copper is <u>less reactive</u> than <u>hydrogen</u> so it doesn't react with dilute acids at all.

MAGNESIUM **ALUMINIUM** **ZINC** **IRON** **COPPER**

1) The more <u>reactive</u> the metal, the <u>faster</u> the reaction will go — very reactive metals (e.g. sodium) react <u>explosively</u>!

2) <u>Copper</u> does <u>not</u> react with dilute acids <u>at all</u> — because it's <u>less</u> reactive than <u>hydrogen</u>.

3) The <u>speed</u> of reaction is indicated by the <u>rate</u> at which the <u>bubbles</u> of hydrogen are given off.

4) The <u>hydrogen</u> is confirmed by the <u>burning splint test</u> giving the notorious 'squeaky pop'.

5) The <u>name</u> of the <u>salt</u> produced depends on which <u>metal</u> is used, and which <u>acid</u> is used:

Hydrochloric Acid Will Always Produce Chloride Salts:

$$2HCl + Mg \rightarrow MgCl_2 + H_2$$ (Magnesium chloride)

$$6HCl + 2Al \rightarrow 2AlCl_3 + 3H_2$$ (Aluminium chloride)

$$2HCl + Zn \rightarrow ZnCl_2 + H_2$$ (Zinc chloride)

Sulfuric Acid Will Always Produce Sulfate Salts:

$$H_2SO_4 + Mg \rightarrow MgSO_4 + H_2$$ (Magnesium sulfate)

$$3H_2SO_4 + 2Al \rightarrow Al_2(SO_4)_3 + 3H_2$$ (Aluminium sulfate)

$$H_2SO_4 + Zn \rightarrow ZnSO_4 + H_2$$ (Zinc sulfate)

Chloride and sulfate salts are generally <u>soluble in water</u>
(the main exceptions are lead chloride, lead sulfate and silver chloride, which are insoluble).

Nitric Acid Produces Nitrate Salts When NEUTRALISED, But...

Nitric acid reacts fine with alkalis, to produce nitrates, but it can play silly devils with metals and produce nitrogen oxides instead, so we'll ignore it here. Chemistry's a real messy subject sometimes, innit.

Nitric acid, tut — there's always one...

Okay, so this stuff isn't exactly a laugh a minute, but at least it's fairly straightforward learning. Metals that are <u>less</u> reactive than <u>hydrogen</u> don't react with acid, and some metals like sodium and potassium are <u>too</u> reactive to mix with acid — your beaker would <u>explode</u>!

Oxides, Hydroxides and Ammonia

Metal Oxides and Metal Hydroxides Are Bases

1) Some metal oxides and metal hydroxides dissolve in water. These soluble compounds are alkalis.
2) Even bases that won't dissolve in water will still react with acids.
3) So, all metal oxides and metal hydroxides react with acids to form a salt and water.

$$Acid + Metal\ Oxide \rightarrow Salt + Water$$

$$Acid + Metal\ Hydroxide \rightarrow Salt + Water$$

(These are neutralisation reactions of course)

The Combination of Metal and Acid Decides the Salt

This isn't exactly exciting but it's pretty easy, so try and get the hang of it:

Hydrochloric acid	+ Copper oxide	→	Copper chloride	+ water
Hydrochloric acid	+ Sodium hydroxide	→	Sodium chloride	+ water
Sulfuric acid	+ Zinc oxide	→	Zinc sulfate	+ water
Sulfuric acid	+ Calcium hydroxide	→	Calcium sulfate	+ water
Nitric acid	+ Magnesium oxide	→	Magnesium nitrate	+ water
Nitric acid	+ Potassium hydroxide	→	Potassium nitrate	+ water

The symbol equations are all pretty much the same. Here are two of them:

$$H_2SO_{4\,(aq)} + ZnO_{(s)} \rightarrow ZnSO_{4\,(aq)} + H_2O_{(l)}$$

$$HNO_{3\,(aq)} + KOH_{(aq)} \rightarrow KNO_{3\,(aq)} + H_2O_{(l)}$$

Ammonia Can Be Neutralised with HNO$_3$ to Make Fertiliser

Ammonia dissolves in water to make an alkaline solution.
When it reacts with nitric acid, you get a neutral salt — ammonium nitrate:

$$NH_{3\,(g)} + HNO_{3\,(aq)} \rightarrow NH_4NO_{3\,(aq)}$$

Ammonia + Nitric acid → Ammonium nitrate

This is a bit different from most neutralisation reactions because there's **NO WATER** produced — just the ammonium salt.

Ammonium nitrate is an especially good fertiliser because it has nitrogen from two sources, the ammonia and the nitric acid. Kind of a double dose. Plants need nitrogen to make proteins.

Magnesium chloride — not so tasty on your chips...

Not the most thrilling of pages, I'm afraid. Just loads of reactions for you to learn. Try doing different combinations of acids and alkalis. Balance them. Cover the page and scribble all the equations down. If you make any mistakes... learn it again, cover it up again, and scribble it all down again.

Making Salts

Most chlorides, sulfates and nitrates are soluble in water (the main exceptions are lead chloride, lead sulfate and silver chloride). Most oxides, hydroxides and carbonates are insoluble in water.

Making Soluble Salts from Insoluble Bases

1) You need to pick the right acid, plus a metal carbonate or metal hydroxide, as long as it's insoluble. You can't use sodium, potassium or ammonium carbonates or hydroxides, as they're soluble (so you can't tell whether the reaction has finished).

2) You add the carbonate or hydroxide to the acid until all the acid is neutralised. (The excess carbonate or hydroxide will just sink to the bottom of the flask when all the acid has reacted.)

3) Then filter out the excess carbonate, and evaporate off the water — and you should be left with a pure, dry salt.

Filtering — to get rid of the excess carbonate or hydroxide.

E.g. you can use copper carbonate and nitric acid to make copper nitrate:

$$CuCO_{3\,(s)} + 2HNO_{3\,(aq)} \longrightarrow Cu(NO_3)_{2\,(aq)} + CO_{2\,(g)} + H_2O_{\,(l)}$$

Making Insoluble Salts — Precipitation Reactions

Just mix an acid and a nitrate — simple as that.

1) If the salt you want to make is insoluble, you can use a precipitation reaction.

2) You just need to pick the right acid and nitrate, then mix them together. E.g. if you want to make lead chloride (which is insoluble), mix hydrochloric acid and lead nitrate.

3) Once the salt has precipitated out (and is lying at the bottom of your flask), all you have to do is filter it from the solution, wash it and then dry it on filter paper.

E.g. $$Pb(NO_3)_{2\,(aq)} + 2HCl_{\,(aq)} \longrightarrow PbCl_{2\,(s)} + 2HNO_{3\,(aq)}$$

4) Precipitation reactions can be used to remove poisonous ions (e.g. lead) from drinking water. Calcium and magnesium ions can also be removed from water this way — they make water "hard", which stops soap lathering properly.

> If both the base and the salt are SOLUBLE, things get a bit trickier. You can't just add an excess of base and filter out what's left — you have to add exactly the right amount of base to just neutralise the acid. You need to use an indicator to show when the reaction's finished. Then repeat using exactly the same volumes of base and acid so the salt isn't contaminated with indicator. All this is quite fiddly.

Making Salts by Displacement

1) If you put a more reactive metal like magnesium into a salt solution of a less reactive metal, like copper sulfate, then the magnesium will take the place of the copper — and make magnesium sulfate.

2) The "kicked-out" (or displaced) metal then coats itself onto the more reactive metal.

3) Once the magnesium has been completely coated with copper, the reaction stops, so this isn't a very practical way to make a salt.

Get two beakers, mix 'em together — job's a good'n...

It's hard to find the precise neutral point using universal indicator. There's quite a wide range of "green"s between blue and yellow. There are more accurate indicators, but you don't need to know about them.

Electrolysis and the Half-Equations

Electrolysis Means "Splitting Up with Electricity"

1) Electrolysis is the breaking down of a substance using electricity.

2) It requires a liquid to conduct the electricity, called the electrolyte.

3) Electrolytes are usually free ions dissolved in water, e.g. dissolved salts, and molten ionic substances.

4) In either case it's the free ions which conduct the electricity and allow the whole thing to work.

5) For an electrical circuit to be complete, there's got to be a flow of electrons. Electrons are taken away from ions at the positive anode and given to other ions at the negative cathode. As ions gain or lose electrons they become atoms or molecules and are released.

NaCl dissolved

Molten NaCl

The Electrolysis of a Salt Solution

When common salt (sodium chloride) is electrolysed, it produces three very useful products.

+ve ions are called CATIONS because they're attracted to the -ve cathode.

Hydrogen is produced at the -ve cathode.

-ve ions are called ANIONS because they're attracted to the +ve anode.

Chlorine is produced at the +ve anode.

1) At the cathode, two hydrogen ions accept two electrons to become one hydrogen molecule.

2) At the anode, two chloride (Cl^-) ions lose their electrons and become one chlorine molecule.

3) NaOH is left in the solution.

The Half-Equations — Make Sure the Electrons Balance

The main thing is to make sure the number of electrons is the same for both half-equations. For the above cell the half-equations are:

Cathode:	$2H^+$	+	$2e^-$	$\rightarrow$	H_2	
Anode:	$2Cl^-$	$\rightarrow$	Cl_2	+	$2e^-$	

Faster shopping at Tesco — use Electrolleys...

Electrolysis is fantastic for removing any unwanted hairs from your body. Great for women with moustaches, or men with hairy backs. And even better for the beauty clinic, as they'll get to charge them a small fortune for the treatment. After all it's a very expensive process...

Chemistry 2(iii) — Using Ions in Solution

Electrolysis of Salt Water

Electrolysis of Salt gives Hydrogen, Chlorine and NaOH

Concentrated brine (sodium chloride solution) is electrolysed industrially using a setup a bit like this one. There are three useful products:

a) Hydrogen gas is given off at the cathode.

b) Chlorine gas is given off at the anode.

c) Sodium hydroxide is left in solution.

These are collected, and then used in all sorts of industries to make various products as detailed below.

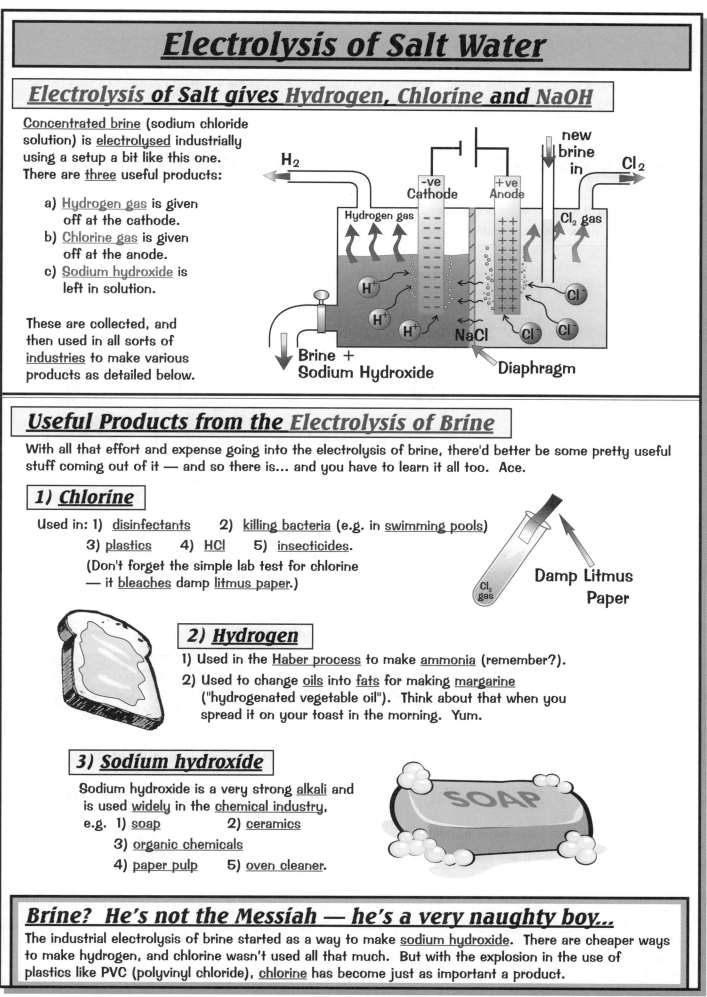

Useful Products from the Electrolysis of Brine

With all that effort and expense going into the electrolysis of brine, there'd better be some pretty useful stuff coming out of it — and so there is... and you have to learn it all too. Ace.

1) Chlorine

Used in: 1) disinfectants   2) killing bacteria (e.g. in swimming pools)

3) plastics   4) HCl   5) insecticides.

(Don't forget the simple lab test for chlorine — it bleaches damp litmus paper.)

Damp Litmus Paper

2) Hydrogen

1) Used in the Haber process to make ammonia (remember?).

2) Used to change oils into fats for making margarine ("hydrogenated vegetable oil"). Think about that when you spread it on your toast in the morning. Yum.

3) Sodium hydroxide

Sodium hydroxide is a very strong alkali and is used widely in the chemical industry,

e.g. 1) soap   2) ceramics
3) organic chemicals
4) paper pulp   5) oven cleaner.

Brine? He's not the Messiah — he's a very naughty boy...

The industrial electrolysis of brine started as a way to make sodium hydroxide. There are cheaper ways to make hydrogen, and chlorine wasn't used all that much. But with the explosion in the use of plastics like PVC (polyvinyl chloride), chlorine has become just as important a product.

Purifying Copper by Electrolysis

1) Copper is a very unreactive metal. Not only is it below carbon in the reactivity series, it's also below hydrogen, which means that copper doesn't even react with water.

2) So copper can be obtained very easily from its ore by reduction with carbon.

Very Pure Copper is Needed for Electrical Conductors

1) The copper produced by reduction isn't pure enough for use in electrical conductors.

2) The purer it is, the better it conducts.

3) Electrolysis is used to obtain very pure copper.

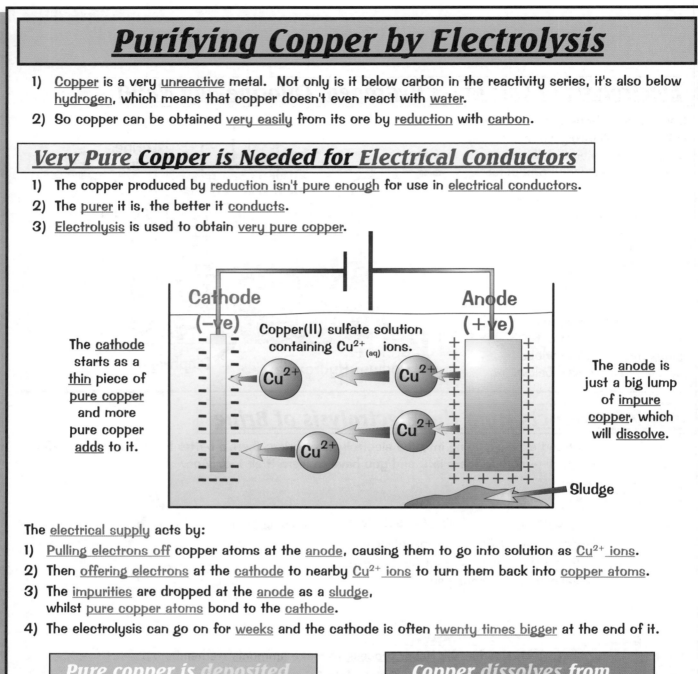

The cathode starts as a thin piece of pure copper and more pure copper adds to it.

Cathode (–ve)

Copper(II) sulfate solution containing $Cu^{2+}_{(aq)}$ ions.

Anode (+ve)

The anode is just a big lump of impure copper, which will dissolve.

Sludge

The electrical supply acts by:

1) Pulling electrons off copper atoms at the anode, causing them to go into solution as Cu^{2+} ions.

2) Then offering electrons at the cathode to nearby Cu^{2+} ions to turn them back into copper atoms.

3) The impurities are dropped at the anode as a sludge, whilst pure copper atoms bond to the cathode.

4) The electrolysis can go on for weeks and the cathode is often twenty times bigger at the end of it.

Pure copper is deposited on the pure cathode (–ve)	*Copper dissolves from the impure anode (+ve)*

The reaction at the cathode is:

$$Cu^{2+}_{(aq)} + 2e^- \rightarrow Cu_{(s)}$$

The copper ions have been converted to copper atoms by gaining electrons.

The reaction at the anode is:

$$Cu_{(s)} \rightarrow Cu^{2+}_{(aq)} + 2e^-$$

Copper atoms have been converted into copper ions by losing electrons.

You do the same thing to purify recycled copper...

Impure copper, like the stuff you get from reducing the ore or from recycling, is okay for most things. It's fine for pipes and pans and decorative stuff. It's only copper in wires and circuit boards that needs to be this pure. Which is handy, because electrolysis is really expensive and uses up lots of energy.

Revision Summary for Chemistry 2(iii)

There are two bits to this section — making salts (by various methods) and electrolysis. Don't try to remember all the individual reactions, just learn the rules. The salt produced depends on the acid and the metal. Have a go at these questions and see how much you can remember.

1) Describe fully the colour of universal indicator for every pH value from 0 to 14.

2) What type of ions are always present in a) acids and b) alkalis?

3) What is neutralisation? Write down the general equation for neutralisation in terms of ions.

4) Give three real-life examples of neutralisation reactions.

5) What is the general equation for reacting an acid with a metal?

6) Which metal(s) don't react at all with acid?

7) Why would you not make a potassium compound by reacting potassium metal with an acid?

8) What type of salts do hydrochloric acid and sulfuric acid produce?

9)* Name the salts formed and write balanced symbol equations for the following reactions:
   a) hydrochloric acid with: i) magnesium, ii) aluminium and iii) zinc,
   b) sulfuric acid with: i) magnesium, ii) aluminium and iii) zinc.

10) What type of reaction is "acid + metal oxide", or "acid + metal hydroxide"?

11) Suggest a suitable acid and a suitable metal oxide/hydroxide to mix to form the following salts.
   a) copper chloride   b) calcium nitrate   c) zinc sulfate
   d) magnesium nitrate   e) sodium sulfate   f) potassium chloride

12) Write a balanced symbol equation for the reaction between ammonia and nitric acid. What is the product of this reaction useful for? Explain why.

13) Iron chloride can made by mixing iron hydroxide (an insoluble base) with hydrochloric acid. Describe the method you would use to produce pure, solid iron chloride in the lab.

14) Describe a practical use of precipitation reactions.

15) How can you tell when a neutralisation reaction is complete if both the base and the salt are soluble in water?

16) What is electrolysis? Explain why only liquids can be electrolysed.

17) What are positive ions called in electrolysis? Why?

18) Write balanced half-equations for the reactions at the anode and the cathode during the electrolysis of sodium chloride solution.

19) Draw a detailed diagram showing how sodium chloride solution (brine) is electrolysed in industry.

20) Give uses for the three products from the electrolysis of brine.

21) Describe the process of purifying copper by electrolysis.

* Answers at the bottom of the page

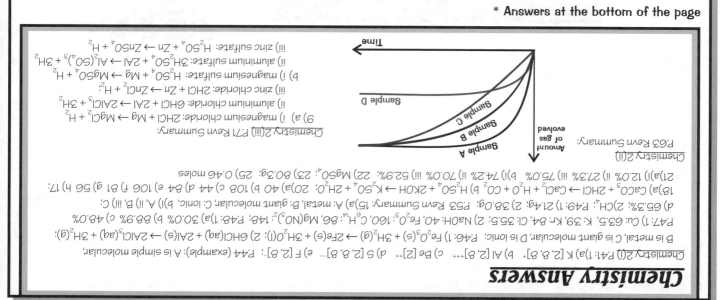

Chemistry 2(i): P41:1)a) K [2,8,8]ᐩ b) Al [2,8]ᐩᐩᐩ c) Be [2]ᐩᐩ d) S [2,8,8]⁻ e) F [2,8]⁻. P44 (example): A is simple molecular. B is metal. C is giant molecular. D is ionic. P46:1) $Fe_2O_3(s) + 3H_2(g) \rightarrow 2Fe(s) + 3H_2O(l)$; 2) $6HCl(aq) + 2Al(s) \rightarrow 2AlCl_3(aq) + 3H_2(g)$; P47:1) Cu: 63.5, K: 39, Kr: 84, Cl: 35.5. 2) NaOH: 40, Fe_2O_3: 160, C_6H_{14}: 86, $Mg(NO_3)_2$: 148. P48:1)a) 30.0% b) 88.9% c) 48.0% d) 65.3%. 2) CH_4; 2) 21.4g; 2) 38.0g. P53 Revn Summry:15)a) A: metal, B: giant molecular, C: ionic, b)i) A, ii) B, iii) C. 18)a) $CaCO_3 \rightarrow CaO + CO_2$; b) $H_2SO_4 + 2KOH \rightarrow K_2SO_4 + 2H_2O$; 20)a) 40 b) 108 c) 44 d) 84 e) 106 f) 81 g) 56 h) 17. 21)a)i) 12.0% ii) 27.3% iii) 75.0% b)i) 74.2% ii) 70.0% iii) 52.9% 22) $MgSO_4$; 23) 80.3g. 25) 0.46 moles

Chemistry 2(iii): P63 Revn Summary:

Chemistry 2(iii): P71 Revn Summary:
9) a) i) magnesium chloride: $2HCl + Mg \rightarrow MgCl_2 + H_2$;
ii) aluminium chloride: $6HCl + 2Al \rightarrow 2AlCl_3 + 3H_2$;
iii) zinc chloride: $2HCl + Zn \rightarrow ZnCl_2 + H_2$;
b) i) magnesium sulfate: $H_2SO_4 + Mg \rightarrow MgSO_4 + H_2$;
ii) aluminium sulfate: $3H_2SO_4 + 2Al \rightarrow Al_2(SO_4)_3 + 3H_2$;
iii) zinc sulfate: $H_2SO_4 + Zn \rightarrow ZnSO_4 + H_2$.

Chemistry Answers

Chemistry 2(iii) — Using Ions in Solution

Velocity and Acceleration

Speed and Velocity are Both How Fast You're Going

Speed and velocity are both measured in <u>m/s</u> (or km/h or mph). They both simply say <u>how fast</u> you're going, but there's a <u>subtle difference</u> between them which <u>you need to know</u>:

> <u>Speed</u> is just <u>how fast</u> you're going (e.g. 30 mph or 20 m/s) with no regard to the direction.
> <u>Velocity</u> however must <u>also</u> have the <u>direction</u> specified, e.g. 30 mph north or 20 m/s, 060°.

Seems kinda fussy I know, but they expect you to remember that distinction, so there you go.

Speed, Distance and Time — the Formula:

$$\text{Speed} = \frac{\text{Distance}}{\text{Time}}$$

You really ought to get <u>pretty slick</u> with this <u>very easy formula</u>.
As usual the <u>formula triangle</u> version makes it all a bit of a <u>breeze</u>.
You just need to try and think up some interesting word for remembering the <u>order</u> of the <u>letters</u> in the triangle, s ^{d} t. Errm... sedit, perhaps... well, you think up your own.

<u>Example:</u> A cat skulks 20 m in 35 s. Find: a) its speed,   b) how long it takes to skulk 75 m.
<u>Answer:</u>   Using the formula triangle: a) s = d/t = 20/35 = <u>0.57 m/s</u>
                                          b) t = d/s = 75/0.57 = 131 s = <u>2 min 11 s</u>

A lot of the time we tend to use the words "speed" and "velocity" interchangeably.
For example, to calculate velocity you'd just use the above formula for speed instead.

Acceleration is How Quickly You're Speeding Up

Acceleration is definitely <u>not</u> the same as <u>velocity</u> or <u>speed</u>.
   Every time you read or write the word <u>acceleration</u>, remind yourself: "<u>acceleration</u> is <u>completely</u> <u>different</u> from <u>velocity</u>. Acceleration is how <u>quickly</u> the velocity is <u>changing</u>."
Velocity is a simple idea. Acceleration is altogether more <u>subtle</u>, which is why it's <u>confusing</u>.

Acceleration — the Formula:

$$\text{Acceleration} = \frac{\text{Change in Velocity}}{\text{Time Taken}}$$

Well, it's <u>just another formula</u>. Just like all the others. Three things in a <u>formula triangle</u>.
Mind you, there are <u>two</u> tricky things with this one. First there's the "Δv", which means working out the "<u>change in velocity</u>", as shown in the example below, rather than just putting a <u>simple value</u> for speed or velocity in. Secondly there's the <u>units</u> of acceleration, which are m/s².
Not m/s, which is <u>velocity</u>, but m/s². Got it? No? Let's try once more: <u>Not m/s, but m/s²</u>.

<u>Example:</u> A skulking cat accelerates steadily from 2 m/s to 6 m/s in 5.6 s. Find its acceleration.
<u>Answer:</u>   Using the formula triangle: a = Δv/t = (6 – 2) / 5.6 = 4 ÷ 5.6 = <u>0.71 m/s²</u>
            All pretty basic stuff I'd say.

Don't speed through this page — learn it properly...

<u>Speed cameras</u> measure the speed of motorists using two photos, taken a fraction of a second apart.
Lines painted on the road help them work out <u>how far</u> the car travelled between the photos — and so <u>how fast</u> it was going. And the photos always have the car's number plate in them. Clever, eh?

D-T and V-T Graphs

Make sure you learn all these details real good. Make sure you can <u>distinguish</u> between the two, too.

Distance-Time Graphs

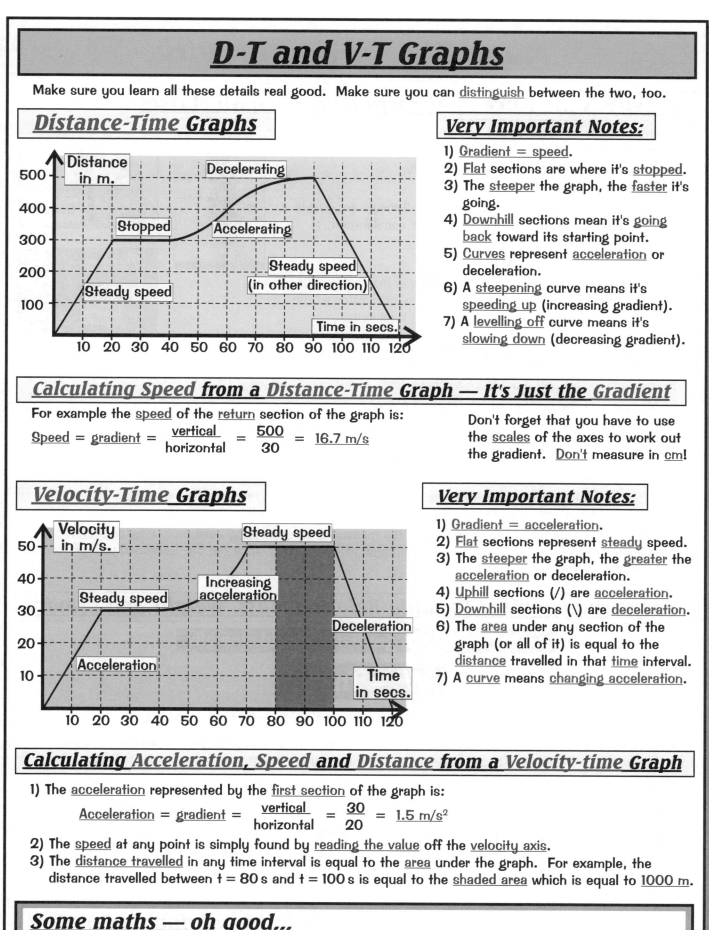

Very Important Notes:

1) <u>Gradient = speed</u>.
2) <u>Flat</u> sections are where it's <u>stopped</u>.
3) The <u>steeper</u> the graph, the <u>faster</u> it's going.
4) <u>Downhill</u> sections mean it's <u>going back</u> toward its starting point.
5) <u>Curves</u> represent <u>acceleration</u> or deceleration.
6) A <u>steepening</u> curve means it's <u>speeding up</u> (increasing gradient).
7) A <u>levelling off</u> curve means it's <u>slowing down</u> (decreasing gradient).

Calculating Speed from a Distance-Time Graph — It's Just the Gradient

For example the <u>speed</u> of the <u>return</u> section of the graph is:

$$\text{Speed} = \text{gradient} = \frac{\text{vertical}}{\text{horizontal}} = \frac{500}{30} = \underline{16.7 \text{ m/s}}$$

Don't forget that you have to use the <u>scales</u> of the axes to work out the gradient. <u>Don't</u> measure in <u>cm</u>!

Velocity-Time Graphs

Very Important Notes:

1) <u>Gradient = acceleration</u>.
2) <u>Flat</u> sections represent <u>steady</u> speed.
3) The <u>steeper</u> the graph, the <u>greater</u> the <u>acceleration</u> or deceleration.
4) <u>Uphill</u> sections (/) are <u>acceleration</u>.
5) <u>Downhill</u> sections (\) are <u>deceleration</u>.
6) The <u>area</u> under any section of the graph (or all of it) is equal to the <u>distance</u> travelled in that <u>time</u> interval.
7) A <u>curve</u> means <u>changing acceleration</u>.

Calculating Acceleration, Speed and Distance from a Velocity-time Graph

1) The <u>acceleration</u> represented by the <u>first section</u> of the graph is:

$$\text{Acceleration} = \text{gradient} = \frac{\text{vertical}}{\text{horizontal}} = \frac{30}{20} = \underline{1.5 \text{ m/s}^2}$$

2) The <u>speed</u> at any point is simply found by <u>reading the value</u> off the <u>velocity axis</u>.
3) The <u>distance travelled</u> in any time interval is equal to the <u>area</u> under the graph. For example, the distance travelled between $t = 80$ s and $t = 100$ s is equal to the <u>shaded area</u> which is equal to <u>1000 m</u>.

Some maths — oh good...

The tricky thing about these two types of graph is that they can look pretty much the same but represent totally different kinds of motion. Make sure you learn all the numbered points, and whenever you're reading a motion graph, check the axis labels carefully so you know which type of graph it is.

Physics 2(i) — Forces and Motion

Mass, Weight and Gravity

Gravity is the Force of Attraction Between All Masses

Gravity attracts all masses, but you only notice it when one of the masses is really really big, e.g. a planet. Anything near a planet or star is attracted to it very strongly.

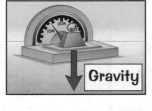

This has three important effects:

1) On the surface of a planet, it makes all things accelerate towards the ground (all with the same acceleration, g, which is about 10 m/s^2 on Earth).

2) It gives everything a weight.

3) It keeps planets, moons and satellites in their orbits. The orbit is a balance between the forward motion of the object and the force of gravity pulling it inwards.

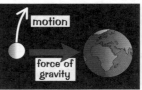

Weight and Mass are Not the Same

To understand this you must learn all these facts about mass and weight:

1) Mass is just the amount of 'stuff' in an object. For any given object this will have the same value anywhere in the Universe.

2) Weight is caused by the pull of gravity. In most questions the weight of an object is just the force of gravity pulling it towards the centre of the Earth.

3) An object has the same mass whether it's on Earth or on the Moon — but its weight will be different. A 1 kg mass will weigh less on the Moon (about 1.6 N) than it does on Earth (about 10 N), simply because the force of gravity pulling on it is less.

4) Weight is a force measured in newtons. It's measured using a spring balance or newton meter. Mass is not a force. It's measured in kilograms with a mass balance (an old-fashioned pair of balancing scales).

The Very Important Formula Relating Mass, Weight and Gravity

weight = mass × gravitational field strength

$$W = m \times g$$

1) Remember, weight and mass are not the same. Mass is in kg, weight is in newtons.

2) The letter "g" represents the strength of the gravity and its value is different for different planets. On Earth g ≈ 10 N/kg. On the Moon, where the gravity is weaker, g is only about 1.6 N/kg.

3) This formula is hideously easy to use:

Example: What is the weight, in newtons, of a 5 kg mass, both on Earth and on the Moon?

Answer: "W = m × g". On Earth: W = 5 × 10 = 50 N   (The weight of the 5 kg mass is 50 N.)
On the Moon: W = 5 × 1.6 = 8 N   (The weight of the 5 kg mass is 8 N.)

See what I mean. Hideously easy — as long as you've learnt what all the letters mean.

Learn about gravity now — no point in "weighting" around...

Often the only way to "understand" something is to learn all the facts about it. And I certainly think that's true here. "Understanding" the difference between mass and weight is no more than learning all the facts about them. When you've learnt all those facts properly, you'll understand it.

The Three Laws of Motion

Around about the time of the Great Plague in the 1660s, a chap called <u>Isaac Newton</u> worked out the <u>Three Laws of Motion</u>. At first they might seem kind of obscure or irrelevant, but to be perfectly blunt, if you can't understand these <u>three simple laws</u> then you'll never understand <u>forces and motion</u>:

First Law — <u>Balanced Forces</u> Mean <u>No Change in Velocity</u>

So long as the forces on an object are all <u>balanced</u>, then it'll just <u>stay still</u>, or else if it's already moving it'll just carry on at the <u>same velocity</u> — so long as the forces are all <u>balanced</u>.

1) When a train or car or bus or anything else is <u>moving</u> at a <u>constant velocity</u> then the <u>forces</u> on it must all be <u>balanced</u>.

2) Never let yourself entertain the <u>ridiculous idea</u> that things need a constant overall force to <u>keep</u> them moving — NO NO NO NO NO NO!

3) To keep going at a <u>steady speed</u>, there must be <u>zero resultant force</u> — and don't you forget it.

Second Law — <u>A Resultant Force</u> Means <u>Acceleration</u>

If there is an <u>unbalanced force</u>, then the object will <u>accelerate</u> in that direction.

1) An <u>unbalanced</u> force will always produce <u>acceleration</u> (or deceleration).

2) This "<u>acceleration</u>" can take <u>five</u> different forms: <u>Starting</u>, <u>stopping</u>, <u>speeding up</u>, <u>slowing down</u> and <u>changing direction</u>.

3) On a force diagram, the <u>arrows</u> will be <u>unequal</u>:

<u>Don't ever say</u>: "If something's moving there must be an overall resultant force acting on it".

Not so. If there's an <u>overall</u> force it will always <u>accelerate</u>.
You get <u>steady</u> speed from <u>balanced</u> forces. I wonder how many times I need to say that same thing before you remember it?

Three Points <u>Which Should Be</u> <u>Obvious:</u>

1) The bigger the <u>force</u>, the <u>greater</u> the <u>acceleration</u> or <u>deceleration</u>.

2) The bigger the <u>mass</u>, the <u>smaller the acceleration</u>.

3) To get a <u>big</u> mass to accelerate <u>as fast</u> as a <u>small</u> mass it needs a <u>bigger</u> force. Just think about pushing <u>heavy</u> trolleys and it should all seem <u>fairly obvious</u>, I would hope.

<u>The Overall</u> Unbalanced Force <u>is Often Called the</u> Resultant Force

Any <u>resultant force</u> will produce <u>acceleration</u>, and this is the <u>formula</u> for it:

$$F = ma \quad \text{or} \quad a = F/m$$

m = mass, a = acceleration, F is always the <u>resultant force</u>.

The Three Laws of Motion

Resultant Force is Real Important — Especially for "F = ma"

The notion of <u>resultant force</u> is a really important one for you to get your head round.
It's not especially tricky, it's just that it seems to get kind of <u>ignored</u>.
In most <u>real</u> situations there are at least <u>two forces</u> acting on an object along any direction.
The <u>overall</u> effect of these forces will decide the <u>motion</u> of the object — whether it will <u>accelerate</u>,
<u>decelerate</u> or stay at a <u>steady speed</u>. If the forces all point along the same direction, the "<u>overall effect</u>"
is found by just <u>adding or subtracting</u> them. The overall force you get is called the <u>resultant force</u>.
And when you use the <u>formula</u> "<u>F = ma</u>", F must always be the <u>resultant force</u>.

<u>Example:</u>   A car of mass of 1750 kg has an engine which provides a driving force of 5200 N.
   At 70 mph the drag force acting on the car is 5150 N.
   Find its acceleration a) when first setting off from rest b) at 70 mph.

<u>ANSWER:</u> 1) First draw a force diagram for both cases (no need to show the vertical forces):

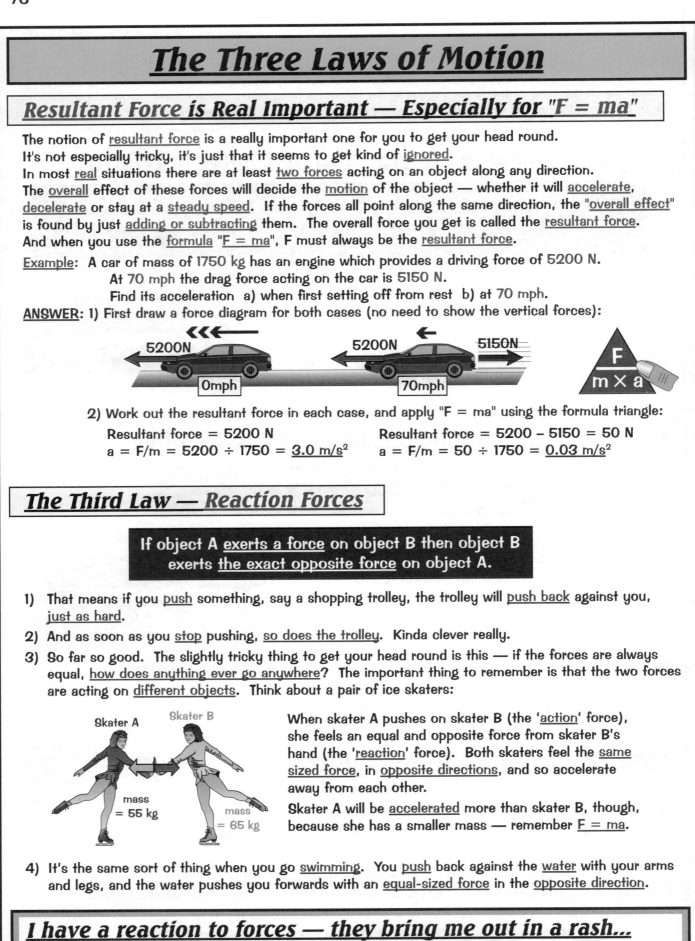

2) Work out the resultant force in each case, and apply "F = ma" using the formula triangle:

Resultant force = 5200 N   Resultant force = 5200 – 5150 = 50 N
a = F/m = 5200 ÷ 1750 = <u>3.0 m/s^2</u>   a = F/m = 50 ÷ 1750 = <u>0.03 m/s^2</u>

The Third Law — Reaction Forces

> If object A <u>exerts a force</u> on object B then object B
> exerts <u>the exact opposite force</u> on object A.

1) That means if you <u>push</u> something, say a shopping trolley, the trolley will <u>push back</u> against you,
<u>just as hard</u>.

2) And as soon as you <u>stop</u> pushing, <u>so does the trolley</u>. Kinda clever really.

3) So far so good. The slightly tricky thing to get your head round is this — if the forces are always
equal, <u>how does anything ever go anywhere</u>? The important thing to remember is that the two forces
are acting on <u>different objects</u>. Think about a pair of ice skaters:

When skater A pushes on skater B (the '<u>action</u>' force),
she feels an equal and opposite force from skater B's
hand (the '<u>reaction</u>' force). Both skaters feel the <u>same
sized force</u>, in <u>opposite directions</u>, and so accelerate
away from each other.

Skater A will be <u>accelerated</u> more than skater B, though,
because she has a smaller mass — remember <u>F = ma</u>.

4) It's the same sort of thing when you go <u>swimming</u>. You <u>push</u> back against the <u>water</u> with your arms
and legs, and the water pushes you forwards with an <u>equal-sized force</u> in the <u>opposite direction</u>.

I have a reaction to forces — they bring me out in a rash...

This is the real deal. Like... proper Physics. It was <u>pretty fantastic</u> at the time — suddenly people
understood how forces and motion worked, they could work out the <u>orbits of planets</u> and everything.
Inspired? No? Shame. Learn them anyway — you're really going to struggle in the exam if you don't.

Drag & Terminal Velocity

1) Friction *is Always There to* Slow Things Down

1) If an object has <u>no force</u> propelling it along it will always <u>slow down and stop</u> because of <u>friction</u> (unless you're in space where there's nothing to rub against).
2) Friction always acts in the <u>opposite</u> direction to movement.
3) To travel at a <u>steady</u> speed, the driving force needs to <u>balance</u> the frictional forces.
4) You get friction between <u>two surfaces</u> in contact, or when an object passes <u>through a fluid</u> (drag).

> **RESISTANCE OR "DRAG" FROM FLUIDS** (air or liquid)
>
> The most important factor <u>by far</u> in <u>reducing drag</u> in fluids is keeping the shape of the object <u>streamlined</u>, like fish bodies or boat hulls or bird wings/bodies. The <u>opposite</u> extreme is a <u>parachute</u> which is about as <u>high drag</u> as you can get — which is, of course, <u>the whole idea</u>.

2) *Drag Increases as the Speed Increases*

Resistance from fluids always <u>increases with speed</u>.

A car has <u>much more</u> friction to <u>work against</u> when travelling at <u>70 mph</u> compared to <u>30 mph</u>. So at 70 mph the engine has to work <u>much harder</u> just to maintain a <u>steady speed</u>.

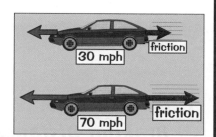

Cars and Free-Fallers Both Reach a Terminal Velocity

When cars and free-falling objects first <u>set off</u> they have <u>much more</u> force <u>accelerating</u> them than <u>resistance</u> slowing them down. As the <u>speed</u> increases the resistance <u>builds up</u>. This gradually <u>reduces</u> the <u>acceleration</u> until eventually the <u>resistance force</u> is <u>equal</u> to the <u>accelerating force</u> and then it won't accelerate any more. It will have reached its maximum speed or <u>terminal velocity</u>.

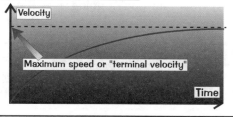

The Terminal Velocity of Falling Objects Depends on their Shape and Area

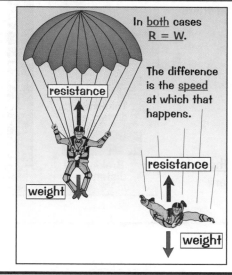

In <u>both</u> cases R = W.

The difference is the <u>speed</u> at which that happens.

The <u>accelerating force</u> acting on <u>all</u> falling objects is <u>gravity</u> and it would make them all fall at the <u>same</u> rate, if it wasn't for <u>air resistance</u>. This means that on the Moon, where there's <u>no air</u>, hamsters and feathers dropped simultaneously will hit the ground <u>together</u>. However, on Earth, <u>air resistance</u> causes things to fall at <u>different</u> speeds, and the <u>terminal velocity</u> of any object is determined by its <u>drag</u> in <u>comparison</u> to its <u>weight</u>.
The drag depends on its <u>shape and area</u>.

The most important example is the human <u>skydiver</u>. Without his parachute open he has quite a <u>small</u> area and a force of "<u>W = mg</u>" pulling him down. He reaches a <u>terminal velocity</u> of about <u>120 mph</u>. But with the parachute <u>open</u>, there's much more <u>air resistance</u> (at any given speed) and still only the same force "<u>W = mg</u>" pulling him down. This means his <u>terminal velocity</u> comes right down to about <u>15 mph</u>, which is a <u>safe speed</u> to hit the ground at.

Learning about air resistance — it can be a real drag...

There are a few really important things on this page. 1) When you fall through a fluid, there's a resistance force (drag), 2) drag increases with speed, so 3) you eventually reach terminal velocity.

Stopping Distances

Looking at things simply — if you <u>need to stop</u> in a <u>given distance</u>... then the <u>faster</u> you're going, the <u>bigger braking force</u> you'll need. But in real life it's not quite that simple — if your maximum braking force isn't enough, you'll go further before you stop. There are loads of <u>other factors</u> too...

Many Factors *Affect Your Total Stopping Distance*

The stopping distance of a car is the distance covered in the time between the driver <u>first spotting</u> a hazard and the car coming to a <u>complete stop</u>. They're pretty keen on this for exam questions, so make sure you <u>learn it properly</u>.

The distance it takes to stop a car is divided into the <u>thinking distance</u> and the <u>braking distance</u>.

1) *Thinking Distance*

"The distance the car travels in the time between the driver noticing the hazard and applying the brakes".

It's affected by <u>two main factors</u>:

a) How fast you're going — obviously. Whatever your reaction time, the <u>faster</u> you're going, the <u>further</u> you'll go.

b) How dopey you are — This is affected by <u>tiredness</u>, <u>drugs</u>, <u>alcohol</u>, <u>old age</u>, and a <u>careless</u> blasé attitude.

> The figures below for typical stopping distances are from the Highway Code. It's frightening to see just how far it takes to stop when you're going at 70 mph.

2) *Braking Distance*

"The distance the car travels during its deceleration whilst the brakes are being applied".

It's affected by <u>four main factors</u>:

a) How fast you're going — The <u>faster</u> you're going the <u>further</u> it takes to stop. More details on page 80.

b) How heavily loaded the vehicle is — with the <u>same</u> brakes, <u>a heavily laden</u> vehicle takes <u>longer to stop</u>. A car won't stop as quick when it's full of people and luggage and towing a caravan.

c) How good your brakes are — all brakes must be checked and maintained <u>regularly</u>. Worn or faulty brakes will let you down <u>catastrophically</u> just when you need them the <u>most</u>, i.e. in an <u>emergency</u>.

d) How good the grip is — this depends on <u>three things</u>:

1) <u>road surface</u>, 2) <u>weather</u> conditions, 3) <u>tyres</u>.

<u>Bad visibility</u> can also be a major factor in accidents — lashing rain, thick fog, bright oncoming lights, etc. might mean that a driver <u>doesn't notice</u> a hazard until they're quite close to it — so they have a much shorter distance available to stop in.

So even at <u>30 mph</u>, you should drive no closer than <u>6 or 7 car lengths</u> away from the car in front — just in case. This is why <u>speed limits</u> are so important, and some <u>residential areas</u> are now <u>20 mph zones</u>.

30 mph	50 mph	70 mph
9 m	15 m	21 m
14 m	38 m	75 m
6 car lengths	13 car lengths	24 car lengths

Thinking distance

Braking distance

Stop right there — and learn this page...

Leaves and diesel spills and muck on t'road are <u>serious hazards</u> because they're <u>unexpected</u>. <u>Wet</u> or <u>icy roads</u> are always much more <u>slippy</u> than dry roads, but often you only discover this when you try to <u>brake</u> hard! Tyres should have a minimum <u>tread depth</u> of <u>1.6 mm</u>. This is essential for getting rid of the <u>water</u> in wet conditions. Without <u>tread</u>, a tyre will simply <u>ride</u> on a <u>layer of water</u> and skid <u>very easily</u>. This is called "<u>aquaplaning</u>" and isn't nearly as cool as it sounds.

Work Done

In Physics, "work done" means something special — it's got its own formula and everything.

"Work Done" is Just "Energy Transferred"

When a force moves an object, energy is transferred and work is done.

That statement sounds far more complicated than it needs to. Try this:

1) Whenever something <u>moves</u>, something else is providing some sort of "<u>effort</u>" to move it.

2) The thing putting the <u>effort</u> in needs a <u>supply</u> of energy (like <u>fuel</u> or <u>food</u> or <u>electricity</u> etc.).

3) It then does "<u>work</u>" by <u>moving</u> the object — and one way or another it <u>transfers</u> the energy it receives (as fuel) into <u>other forms</u>.

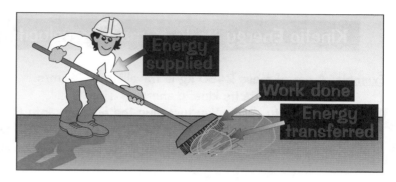

4) Whether this energy is transferred "<u>usefully</u>" (e.g. by <u>lifting a load</u>) or is "<u>wasted</u>" (e.g. lost as <u>heat</u>), you can still say that "<u>work is done</u>". Just like Batman and Bruce Wayne, "<u>work done</u>" and "<u>energy transferred</u>" are indeed "<u>one and the same</u>". (And they're both in <u>joules</u>.)

5) When you do work against <u>friction</u>, most of the energy gets transformed into <u>heat</u>, and usually some into <u>sound</u>. This is generally "<u>wasted</u>" energy.

And Another Formula to Learn...

Work Done = Force × Distance

This formula only works if the force is in exactly the same direction as the movement.

Whether the force is <u>friction</u> or <u>weight</u> or <u>tension in a rope</u>, it's always the same. To find how much <u>work</u> has been <u>done</u> (in joules), you just multiply the <u>force in newtons</u> by the <u>distance moved in metres</u>. Easy as that. I'll show you...

> <u>Example:</u> Some hooligan kids drag an old tractor tyre 5 m over flat ground.
> They pull with a total force of 340 N.
> Find the work done.
>
> <u>Answer:</u> W = F × d = 340 × 5 = <u>1700 J</u>.

Phew — easy peasy isn't it?

Revise work done — what else...

So, work is just energy transferred. Learn the formula then have a go at this question:

1) An electric winch does 2058 J of work in lifting a 42 kg sheep up into the air.
Through what height does the winch raise the sheep? (Take g = 10 m/s^2)

Answer on page 100

Kinetic and Potential Energy

Kinetic Energy is Energy of Movement

Anything which is moving has kinetic energy.
The kinetic energy of something depends both on its mass and speed.
The greater its mass and the faster it's going, the bigger its kinetic energy will be.

There's a slightly tricky formula for it, so you have to concentrate a little bit harder for this one.
But hey, that's life — it can be real tough sometimes:

> ## Kinetic Energy = ½ × mass × velocity²

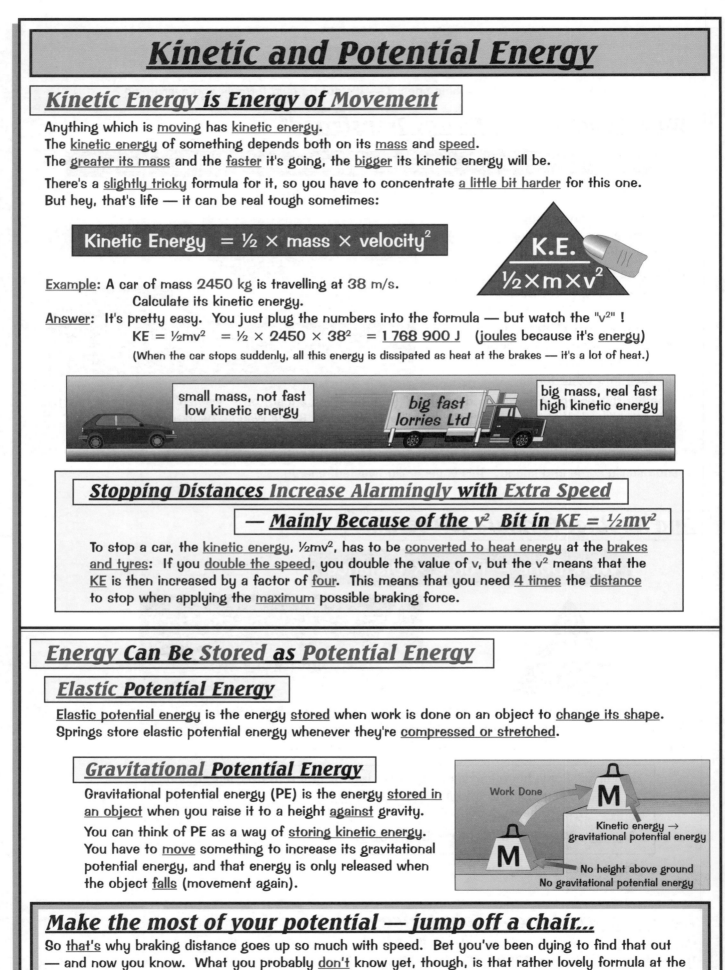

Example: A car of mass 2450 kg is travelling at 38 m/s.
Calculate its kinetic energy.

Answer: It's pretty easy. You just plug the numbers into the formula — but watch the "v²" !
$$KE = \tfrac{1}{2}mv^2 = \tfrac{1}{2} \times 2450 \times 38^2 = \underline{1\,768\,900\ J} \quad \text{(joules because it's energy)}$$
(When the car stops suddenly, all this energy is dissipated as heat at the brakes — it's a lot of heat.)

small mass, not fast
low kinetic energy

big fast lorries Ltd

big mass, real fast
high kinetic energy

Stopping Distances Increase Alarmingly with Extra Speed

— Mainly Because of the v² Bit in KE = ½mv²

To stop a car, the kinetic energy, ½mv², has to be converted to heat energy at the brakes and tyres: If you double the speed, you double the value of v, but the v² means that the KE is then increased by a factor of four. This means that you need 4 times the distance to stop when applying the maximum possible braking force.

Energy Can Be Stored as Potential Energy

Elastic Potential Energy

Elastic potential energy is the energy stored when work is done on an object to change its shape.
Springs store elastic potential energy whenever they're compressed or stretched.

Gravitational Potential Energy

Gravitational potential energy (PE) is the energy stored in an object when you raise it to a height against gravity.

You can think of PE as a way of storing kinetic energy. You have to move something to increase its gravitational potential energy, and that energy is only released when the object falls (movement again).

Work Done

Kinetic energy →
gravitational potential energy

No height above ground
No gravitational potential energy

Make the most of your potential — jump off a chair...

So that's why braking distance goes up so much with speed. Bet you've been dying to find that out — and now you know. What you probably don't know yet, though, is that rather lovely formula at the top of the page. I mean, gosh, it's got more than three letters in it. Get learning.

Momentum and Collisions

A <u>large</u> rugby player running very <u>fast</u> is going to be a lot harder to stop than a scrawny one out for a Sunday afternoon stroll — that's momentum for you.

Momentum = Mass × Velocity

1) The <u>greater</u> the <u>mass</u> of an object and the <u>greater</u> its <u>velocity</u>, the <u>more momentum</u> the object has.

2) Momentum is a <u>vector</u> quantity — it has size <u>and</u> direction (like <u>velocity</u>, but not speed).

$$\frac{\text{momentum}}{\text{mass} \times \text{velocity}}$$

Momentum (kg m/s) = Mass (kg) × Velocity (m/s)

Momentum Before = Momentum After

<u>Momentum is conserved</u> when no external forces act, i.e. the total momentum <u>after</u> is the <u>same</u> as it was <u>before</u>.

Example 1:

Two skaters approach each other, collide and move off together as shown. At what velocity do they move after the collision?

2 m/s — Ed — 80 kg — Before
1.5 m/s — Sue — 60 kg
Velocity (v) = ? — (80+60) kg — After

1) Choose which direction is <u>positive</u>. I'll say "<u>positive</u>" means "<u>to the right</u>".

2) <u>Total momentum before</u> collision
= momentum of Ed + momentum of Sue
= {80 × 2} + {60 × (–1.5)} = <u>70 kg m/s</u>

3) <u>Total momentum after</u> collision
= momentum of Ed and Sue together
= <u>140 × v</u>

4) So 140v = 70, i.e. <u>v = 0.5 m/s to the right</u>

Example 2:

A gun fires a bullet as shown. At what speed does the gun move backwards?

Velocity (v) = ? — 150 m/s — 1 kg — 0.01 kg — After

1) Choose which direction is <u>positive</u>. Again, I reckon "<u>positive</u>" means "<u>to the right</u>".

2) <u>Total momentum before</u> firing
= <u>0 kg m/s</u>

3) <u>Total momentum after</u> firing
= momentum of bullet + momentum of gun
= (0.01 × 150) + (1 × v)
= <u>1.5 + v</u>

This is the gun's <u>recoil</u>.

4) So 1.5 + v = 0, i.e. v = –1.5 m/s
So the gun moves backwards at <u>1.5 m/s</u>.

Forces Cause Changes in Momentum

1) When a <u>force</u> acts on an object, it causes a <u>change</u> in momentum.

$$\text{Force acting (N)} = \frac{\text{Change in momentum (kg m/s)}}{\text{Time taken for change to happen (s)}}$$

2) A <u>larger</u> force means a <u>faster</u> change of momentum (and so a greater <u>acceleration</u>).

3) Likewise, if someone's momentum changes <u>very quickly</u> (like in a <u>car crash</u>), the <u>forces</u> on the body will be very <u>large</u>, and more likely to cause <u>injury</u>.

4) This is why cars are designed to slow people down over a <u>longer time</u> when they have a crash — the longer it takes for a change in <u>momentum</u>, the <u>smaller</u> the <u>force</u>.

<u>CRUMPLE ZONES</u> crumple on impact, <u>increasing the time</u> taken for the car to stop.

<u>SEAT BELTS</u> stretch slightly, <u>increasing the time</u> taken for the wearer to stop. This <u>reduces the forces</u> acting on the chest.

<u>AIR BAGS</u> also slow you down more <u>gradually</u>.

Learn this stuff — it'll only take a moment... um...

Momentum's a pretty fundamental bit of Physics — so make sure you learn it properly. Right then, momentum is always conserved in collisions and explosions when there are no external forces acting. The bit at the bottom of the page is just another way of writing Newton's 2nd law of motion. Learn it.

Revision Summary for Physics 2(i)

Yay — revision summary! I <u>know</u> these are your favourite bits of the book, all those jolly questions. There are lots of formulas and laws and picky little details to learn in this section. So, practise these questions till you can do them all standing on one leg with your arms behind your back whilst being tickled on the nose with a purple ostrich feather. Or something.

1)   What's the difference between speed and velocity? Give an example of each.
2)* Write down the formula for working out speed. Find the speed of a partly chewed mouse which hobbles 3.2 m in 35 s. Find how far he would get in 25 minutes.
3)* A speed camera is set up in a 30 mph (13.4 m/s) zone. It takes two photographs 0.5 s apart. A car travels 6.3 m between the two photographs. Was the car breaking the speed limit?
4)   What is acceleration? What are its units?
5)* Write down the formula for acceleration.
      What's the acceleration of a soggy pea, flicked from rest to a speed of 14 m/s in 0.4 s?
6)   Sketch a typical distance-time graph and point out all the important parts of it.
7)   Explain how to calculate velocity from a distance-time graph.
8)   Sketch a typical velocity-time graph and point out all the important parts of it.
9)   Explain how to find speed, distance and acceleration from a velocity-time graph.
10) What is gravity? List three effects gravity produces.
11) Explain the difference between mass and weight. What units are they measured in?
12) What's the formula for weight? Illustrate it with a worked example of your own.
13) Write down Newton's first law of motion. Illustrate it with a diagram.
14) Write down Newton's second law of motion. Illustrate it with a diagram. What's the formula for it?
15)* A force of 30 N pushes on a trolley of mass 4 kg. What will be its acceleration?
16)* What's the mass of a cat which accelerates at 9.8 m/s² when acted on by a force of 56 N?
17) Explain what "resultant force" is. Illustrate with a diagram.
18)* A skydiver has a mass of 75 kg. At 80 mph, the drag force on the skydiver is 650 N.
      Find the acceleration of the skydiver at 80 mph (take g = 10 N/kg).
19) Write down Newton's third law of motion. Illustrate it with a diagram.
20) What is "terminal velocity"?
21) Describe how friction in fluids is affected by speed.
22) What are the two main factors affecting the terminal velocity of a falling object?
23) What are the two different parts of the overall stopping distance of a car?
24) List the factors which affect each of the two sections of stopping distance.
25)* What's the formula for work done? A crazy dog drags a big branch 12 m over the next-door neighbour's front lawn, pulling with a force of 535 N. How much work was done?
26)* Write down the formula for kinetic energy. Find the KE of a 78 kg sheep moving at 23 m/s.
27) Explain why the stopping distance of a car increases so much with speed.
28) What are elastic potential energy and gravitational potential energy?
29)* Write down the formula for momentum. Find the momentum of a 78 kg sheep falling at 15 m/s.
30) If the total momentum of a system before a collision is zero, what is the total momentum of the system after the collision?
31)* A gymnast (mass 50 kg) jumps off a beam and hits the floor at a speed of 7 m/s.
      She bends her knees and stops moving in 0.5 s. What is the average force acting on her?
32) Explain how air bags, seat belts and crumple zones reduce the risk of serious injury in a car crash.

* Answers on page 100

Physics 2(i) — Forces and Motion

Static Electricity

Static electricity is all about charges which are <u>not</u> free to move. This causes them to build up in one place and it often ends with a <u>spark</u> or a <u>shock</u> when they do finally move.

1) Build-up of Static is Caused by Friction

1) When two <u>insulating</u> materials are <u>rubbed</u> together, electrons will be <u>scraped off one</u> and <u>dumped</u> on the other.

2) This'll leave a <u>positive</u> static charge on one and a <u>negative</u> static charge on the other.

3) <u>Which way</u> the electrons are transferred <u>depends</u> on the <u>two materials</u> involved.

4) Electrically charged objects <u>attract</u> small objects placed near them.
(Try this: rub a balloon on a woolly pully — then put it near tiddly bits of paper and watch them jump.)

5) The classic examples are <u>polythene</u> and <u>acetate</u> rods being rubbed with a <u>cloth duster</u>, as shown in the diagrams.

With the <u>polythene rod</u>, electrons move <u>from the duster</u> to the rod.

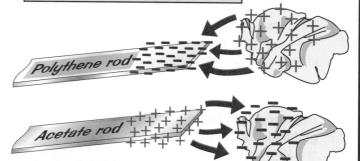

With the <u>acetate rod</u>, electrons move <u>from the rod</u> to the duster.

2) Only Electrons Move — Never the Positive Charges

<u>Watch out for this in exams</u>. Both +ve and −ve electrostatic charges are only ever produced by the movement of <u>electrons</u>. The positive charges <u>definitely do not move</u>! A positive static charge is always caused by electrons <u>moving</u> away elsewhere, as shown above. Don't forget!

A charged conductor can be <u>discharged safely</u> by connecting it to earth with a <u>metal strap</u>. The electrons flow <u>down</u> the strap to the ground if the charge is <u>negative</u> and flow <u>up</u> the strap from the ground if the charge is <u>positive</u>.

The <u>rate of flow</u> of <u>electrical charge</u> is called <u>electric current</u> (see p85).

3) Like Charges Repel, Opposite Charges Attract

This is <u>easy</u> and, I'd have thought, <u>kind of obvious</u>.
Two things with <u>opposite</u> electric charges are <u>attracted</u> to each other.
Two things with the <u>same</u> electric charge will <u>repel</u> each other.
These forces get <u>weaker</u> the <u>further apart</u> the two things are.

4) As Charge Builds Up, So Does the Voltage — Causing Sparks

The greater the <u>charge</u> on an <u>isolated</u> object, the greater the <u>voltage</u> between it and the Earth. If the voltage gets <u>big enough</u> there's a <u>spark</u> which <u>jumps</u> across the gap. High voltage cables can be <u>dangerous</u> for this reason. Big sparks have been known to <u>leap</u> from <u>overhead cables</u> to earth. But not often.

'ZAP!'

Static caravans — where electrons go on holiday...

Static electricity's great fun. You must have tried it — rubbing a balloon against your jumper and trying to get it to stick to the ceiling. It really works... well, sometimes. <u>Bad hair days</u> are caused by static too — it builds up on your hair, so your strands of hair repel each other. Which is nice...

Static Electricity — Examples

They like asking you to give quite detailed examples in exams. Make sure you learn all these details.

Static Electricity Being Helpful:

1) Smoke Precipitators:

Smoke is made up of tiny particles which can be removed with a precipitator. There are several different designs of precipitator — here's a very simple one:

1) As smoke particles reach the bottom of the chimney, they meet a wire grid with a high negative charge and this charges the smoke negatively.

2) The charged smoke particles are attracted to positively charged metal plates. The smoke particles stick together to form larger particles.

3) When heavy enough the particles fall off the plates or are knocked off by a hammer. The dust falls to the bottom of the chimney and can be removed.

4) The gases coming out of the chimney have very little smoke in them.

2) Photocopier:

1) The image plate is positively charged. An image of what you're copying is projected onto it.

2) Whiter bits of the thing you're copying make light fall on the plate and the charge leaks away in those places.

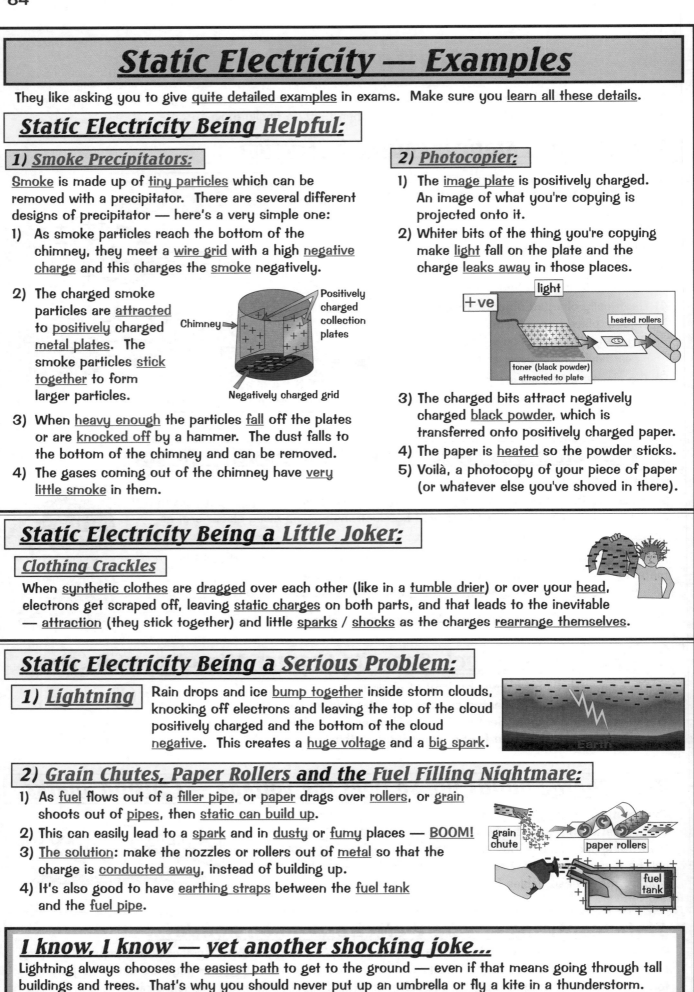

3) The charged bits attract negatively charged black powder, which is transferred onto positively charged paper.

4) The paper is heated so the powder sticks.

5) Voilà, a photocopy of your piece of paper (or whatever else you've shoved in there).

Static Electricity Being a Little Joker:

Clothing Crackles

When synthetic clothes are dragged over each other (like in a tumble drier) or over your head, electrons get scraped off, leaving static charges on both parts, and that leads to the inevitable — attraction (they stick together) and little sparks / shocks as the charges rearrange themselves.

Static Electricity Being a Serious Problem:

1) Lightning

Rain drops and ice bump together inside storm clouds, knocking off electrons and leaving the top of the cloud positively charged and the bottom of the cloud negative. This creates a huge voltage and a big spark.

2) Grain Chutes, Paper Rollers and the Fuel Filling Nightmare:

1) As fuel flows out of a filler pipe, or paper drags over rollers, or grain shoots out of pipes, then static can build up.

2) This can easily lead to a spark and in dusty or fumy places — BOOM!

3) The solution: make the nozzles or rollers out of metal so that the charge is conducted away, instead of building up.

4) It's also good to have earthing straps between the fuel tank and the fuel pipe.

I know, I know — yet another shocking joke...

Lightning always chooses the easiest path to get to the ground — even if that means going through tall buildings and trees. That's why you should never put up an umbrella or fly a kite in a thunderstorm.

Physics 2(ii) — Electricity and the Atom

Circuits — The Basics

Isn't electricity great. Mind you it's pretty bad news if the words don't mean anything to you... Hey, I know — learn them now!

1) Current is the <u>flow</u> of electrons round the circuit.
Current will <u>only flow</u> through a component if there is a <u>voltage</u> across that component.   Unit: ampere, A.

2) Voltage is the <u>driving force</u> that pushes the current round.
Kind of like "<u>electrical pressure</u>".   Unit: volt, V.

3) Resistance is anything in the circuit which <u>slows the flow down</u>.   Unit: ohm, Ω.

4) There's a balance: the <u>voltage</u> is trying to <u>push</u> the current round the circuit, and the <u>resistance</u> is <u>opposing</u> it — the <u>relative sizes</u> of the voltage and resistance decide <u>how big</u> the current will be:

> If you <u>increase the voltage</u> — then <u>more current</u> will flow.
> If you <u>increase the resistance</u> — then <u>less current</u> will flow
> (or <u>more voltage</u> will be needed to keep the <u>same current</u> flowing).

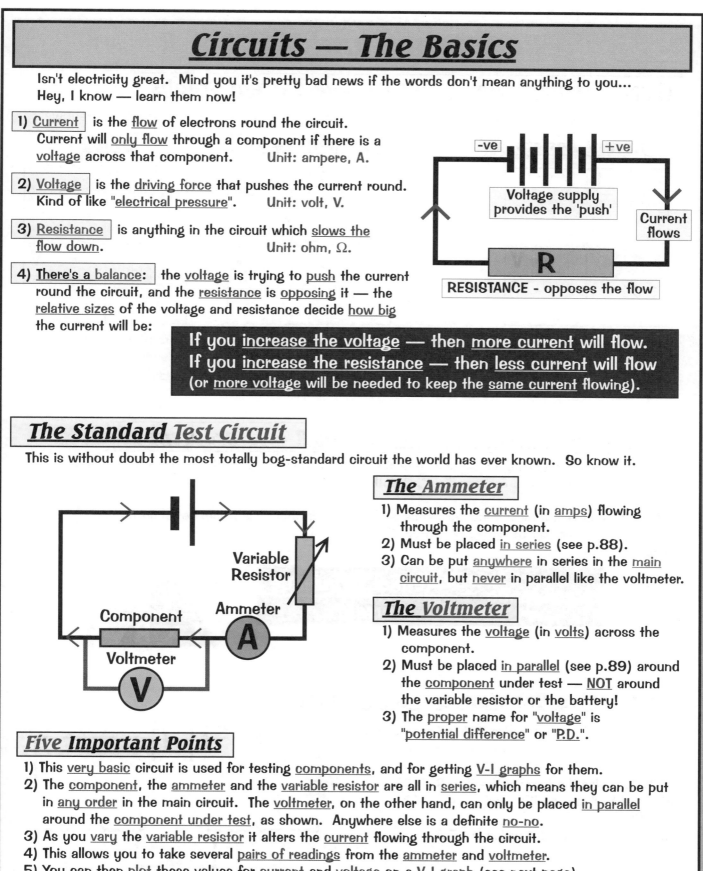

-ve +ve

Voltage supply
provides the 'push'

Current
flows

R

RESISTANCE - opposes the flow

The Standard Test Circuit

This is without doubt the most totally bog-standard circuit the world has ever known. So know it.

Variable
Resistor

Ammeter

Component

Voltmeter

The Ammeter

1) Measures the <u>current</u> (in <u>amps</u>) flowing through the component.
2) Must be placed <u>in series</u> (see p.88).
3) Can be put <u>anywhere</u> in series in the <u>main circuit</u>, but <u>never</u> in parallel like the voltmeter.

The Voltmeter

1) Measures the <u>voltage</u> (in <u>volts</u>) across the component.
2) Must be placed <u>in parallel</u> (see p.89) around the <u>component</u> under test — <u>NOT</u> around the variable resistor or the battery!
3) The <u>proper</u> name for "<u>voltage</u>" is "<u>potential difference</u>" or "<u>P.D.</u>".

Five Important Points

1) This <u>very basic</u> circuit is used for testing <u>components</u>, and for getting <u>V-I graphs</u> for them.
2) The <u>component</u>, the <u>ammeter</u> and the <u>variable resistor</u> are all in <u>series</u>, which means they can be put in <u>any order</u> in the main circuit. The <u>voltmeter</u>, on the other hand, can only be placed <u>in parallel</u> around the <u>component under test</u>, as shown. Anywhere else is a definite <u>no-no</u>.
3) As you <u>vary</u> the <u>variable resistor</u> it alters the <u>current</u> flowing through the circuit.
4) This allows you to take several <u>pairs of readings</u> from the <u>ammeter</u> and <u>voltmeter</u>.
5) You can then <u>plot</u> these values for <u>current</u> and <u>voltage</u> on a <u>V-I graph</u> (see next page).

Measure gymnastics — use a vaultmeter...

The funny thing is — the <u>electrons</u> in circuits actually move from <u>–ve to +ve</u>... but scientists always think of <u>current</u> as flowing from <u>+ve to –ve</u>. Basically it's just because that's how the <u>early physicists</u> thought of it (before they found out about the electrons), and now it's become <u>convention</u>.

Resistance and V = I × R

Three Hideously Important Voltage-Current Graphs

V-I graphs show how the current varies as you change the voltage. Learn these three real well:

Different Resistors

Filament Lamp

Diode

The current through a _resistor_ (at constant temperature) is _proportional to voltage_. _Different resistors_ have different _resistances_, hence the different _slopes_.

As the _temperature_ of the filament _increases_, the _resistance increases_, hence the _curve_.

Current will only flow through a diode _in one direction_, as shown.

Calculating Resistance: R = V/I, (or R = "1/gradient")

For the _straight-line graphs_ the resistance of the component is _steady_ and is equal to the _inverse_ of the _gradient_ of the line, or "_1/gradient_". In other words, the _steeper_ the graph the _lower_ the resistance.

If the graph _curves_, it means the resistance is _changing_. In that case R can be found for any point by taking the _pair of values_ (V, I) from the graph and sticking them in the formula _R = V/I_. Easy.

$$\text{Resistance} = \frac{\text{Potential Difference}}{\text{Current}}$$

$$\frac{V}{I \times R}$$

Calculating Resistance — an Example

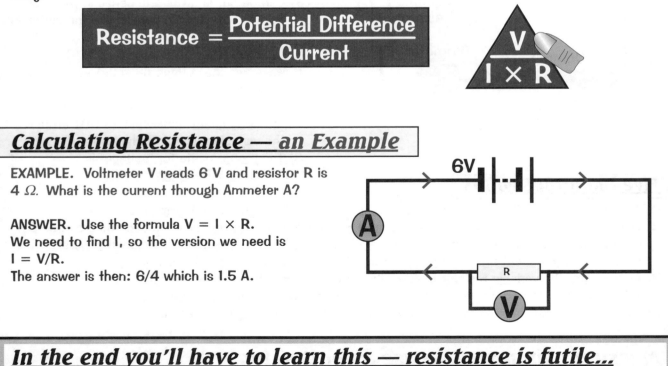

EXAMPLE. Voltmeter V reads 6 V and resistor R is 4 Ω. What is the current through Ammeter A?

ANSWER. Use the formula V = I × R.
We need to find I, so the version we need is
I = V/R.
The answer is then: 6/4 which is 1.5 A.

In the end you'll have to learn this — resistance is futile...

You have to be able to _interpret_ voltage-current graphs for your exam. Remember — the _steeper_ the _slope_, the _lower_ the _resistance_. And you need to know that equation inside out, back to front, upside down and in Swahili. It's the most important equation in electrics, bar none. (PS. I might let you off the Swahili.)

Circuit Symbols and Devices

Circuit Symbols You Should Know:

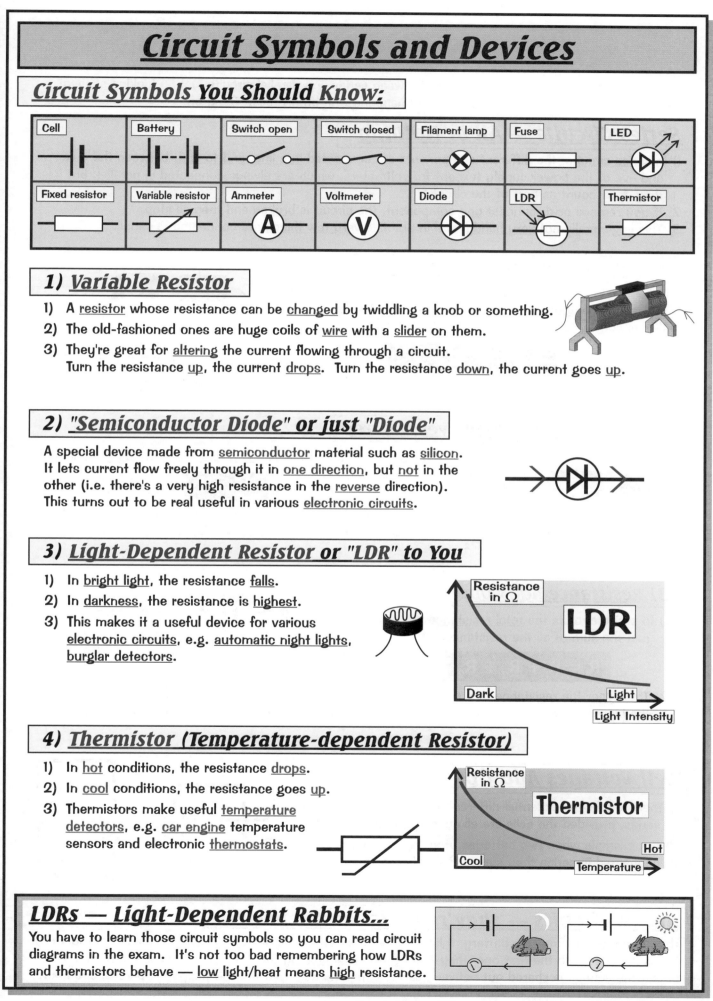

| Cell | Battery | Switch open | Switch closed | Filament lamp | Fuse | LED |
| Fixed resistor | Variable resistor | Ammeter | Voltmeter | Diode | LDR | Thermistor |

1) Variable Resistor

1) A resistor whose resistance can be changed by twiddling a knob or something.
2) The old-fashioned ones are huge coils of wire with a slider on them.
3) They're great for altering the current flowing through a circuit.
   Turn the resistance up, the current drops. Turn the resistance down, the current goes up.

2) "Semiconductor Diode" or just "Diode"

A special device made from semiconductor material such as silicon.
It lets current flow freely through it in one direction, but not in the
other (i.e. there's a very high resistance in the reverse direction).
This turns out to be real useful in various electronic circuits.

3) Light-Dependent Resistor or "LDR" to You

1) In bright light, the resistance falls.
2) In darkness, the resistance is highest.
3) This makes it a useful device for various
   electronic circuits, e.g. automatic night lights,
   burglar detectors.

Resistance in Ω

LDR

Dark   Light

Light Intensity

4) Thermistor (Temperature-dependent Resistor)

1) In hot conditions, the resistance drops.
2) In cool conditions, the resistance goes up.
3) Thermistors make useful temperature
   detectors, e.g. car engine temperature
   sensors and electronic thermostats.

Resistance in Ω

Thermistor

Cool   Hot

Temperature

LDRs — Light-Dependent Rabbits...

You have to learn those circuit symbols so you can read circuit
diagrams in the exam. It's not too bad remembering how LDRs
and thermistors behave — low light/heat means high resistance.

Series Circuits

You need to be able to tell the difference between series and parallel circuits <u>just by looking at them</u>. You also need to know the <u>rules</u> about what happens with both types. Read on.

Series Circuits — All or Nothing

1) In <u>series circuits</u>, the different components are connected <u>in a line</u>, <u>end to end</u>, between the +ve and −ve of the power supply (except for <u>voltmeters</u>, which are always connected <u>in parallel</u>, but they don't count as part of the circuit).
2) If you remove or disconnect <u>one</u> component, the circuit is <u>broken</u> and they all <u>stop</u>.
3) This is generally <u>not very handy</u>, and in practice <u>very few things</u> are connected in series.

1) Potential Difference is Shared:

In series circuits the <u>total P.D.</u> of the <u>supply</u> is <u>shared</u> between the various <u>components</u>. So the <u>voltages</u> round a series circuit <u>always add up</u> to equal the <u>source voltage</u>:

$$V = V_1 + V_2 + V_3$$

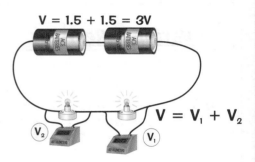

2) Current is the Same Everywhere:

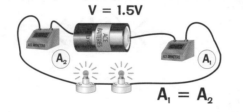

1) In series circuits the <u>same current</u> flows through <u>all parts</u> of the circuit, i.e:

$$A_1 = A_2$$

2) The <u>size</u> of the current is determined by the <u>total P.D.</u> of the cells and the <u>total resistance</u> of the circuit: i.e. $I = V/R$

3) Resistance Adds Up:

1) In series circuits the <u>total resistance</u> is just the <u>sum</u> of all the resistances:

$$R = R_1 + R_2 + R_3$$

2) The <u>bigger</u> the <u>resistance</u> of a component, the bigger its <u>share</u> of the <u>total P.D.</u>

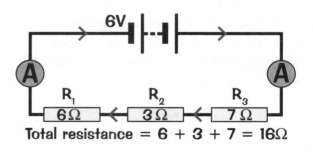

Total resistance = 6 + 3 + 7 = 16Ω

Cell Voltages Add Up:

1) There is a bigger potential difference when more cells are in series, provided the cells are all <u>connected</u> the <u>same way</u>.
2) For example when two batteries of voltage 1.5 V are <u>connected in series</u> they supply 3 V <u>between them</u>.

Series circuits — they're no laughing matter...

If you connect a lamp to a battery, it lights up with a certain brightness. If you then add more identical lamps in series with the first one, they'll all light up <u>less brightly</u> than before. That's because in a series circuit the voltage is <u>shared out</u> between all the components. That doesn't happen in parallel circuits...

Physics 2(ii) — Electricity and the Atom

Parallel Circuits

Parallel circuits are much more <u>sensible</u> than series circuits and so they're much more <u>common</u> in <u>real life</u>. All the electrics in your house will be wired in parallel circuits.

Parallel Circuits — Independence and Isolation

1) In <u>parallel circuits</u>, each component is <u>separately</u> connected to the +ve and –ve of the <u>supply</u>.

2) If you remove or disconnect <u>one</u> of them, it will <u>hardly affect</u> the others at all.

3) This is <u>obviously</u> how <u>most</u> things must be connected, for example in <u>cars</u> and in <u>household</u> <u>electrics</u>. You have to be able to switch everything on and off <u>separately</u>.

1) P.D. is the Same Across All Components:

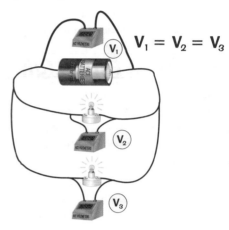

1) In parallel circuits <u>all</u> components get the <u>full source P.D.</u>, so the voltage is the <u>same</u> across all components:

$$V_1 = V_2 = V_3$$

2) This means that <u>identical bulbs</u> connected in parallel will all be at the <u>same brightness</u>.

$V_1 = V_2 = V_3$

2) Current is Shared Between Branches:

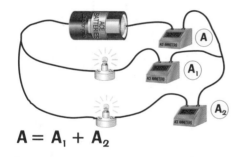

$$A = A_1 + A_2$$

1) In parallel circuits the <u>total current</u> flowing around the circuit is equal to the <u>total</u> of all the currents in the <u>separate branches</u>.

$$A = A_1 + A_2 + A_3$$

2) In a parallel circuit, there are <u>junctions</u> where the current either <u>splits</u> or <u>rejoins</u>. The total current going <u>into</u> a junction has to equal the total current <u>leaving</u>.

3) If two <u>identical components</u> are connected in parallel then the <u>same current</u> will flow through each component.

3) Resistance Is Tricky:

1) The <u>current</u> through each component depends on its <u>resistance</u>. The <u>lower</u> the resistance, the <u>bigger</u> the current that'll flow through it.

2) The <u>total resistance</u> of the circuit is <u>tricky to work out</u>, but it's always <u>less</u> than that of the branch with the <u>smallest</u> resistance.

Voltmeters and Ammeters Are Exceptions to the Rule:

1) Ammeters and voltmeters are <u>exceptions</u> to the series and parallel rules.

2) Ammeters are <u>always</u> connected in <u>series</u> even in a parallel circuit.

3) Voltmeters are <u>always</u> connected in <u>parallel with a component</u> even in a series circuit.

A current shared — is a current halved...*

Parallel circuits might look a bit scarier than series ones, but they're much more useful — and you don't have to learn as many equations for them (yay!). Remember: each branch has the <u>same</u> <u>voltage</u> across it, and the overall resistance is lower than that of the least resistant branch.

* Conditions may apply. CGP takes no
responsibility for the accuracy of this proverb.

Physics 2(ii) — Electricity and the Atom

Series and Parallel Circuits — Examples

Example on Series Circuits

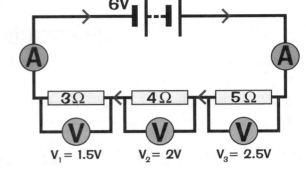

Voltages add to equal the source voltage:
1.5 + 2 + 2.5 = 6 V

Total resistance is the sum of the resistances in the circuit: 3 + 4 + 5 = 12 ohms

Current flowing through all parts of the circuit
= V/R = 6/12 = 0.5 A

(If an extra cell was added of voltage 3 V then the voltage across each resistor would increase.)

Christmas Fairy Lights are Often Wired in Series

Christmas fairy lights are about the only real-life example of things connected in series, and we all know what a pain they are when the whole lot go out just because one of the bulbs is slightly dicky. The only advantage is that the bulbs can be very small because the total 230 V is shared out between them, so each bulb only has a small voltage across it.

Example on Parallel Circuits

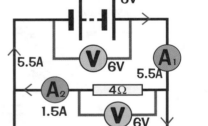
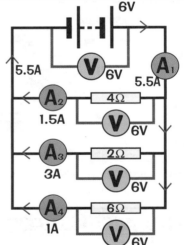

1) The voltage across each resistor in the circuit is the same as the supply voltage. Each voltmeter will read 6 V.

2) The current through each resistor will be different because they have different values of resistance.

3) The current through the battery is the same as the sum of the other currents in the branches.
   i.e. $A_1 = A_2 + A_3 + A_4 \Rightarrow A_1 = 1.5 + 3 + 1 = 5.5$ A

4) The total resistance in the whole circuit is less than the lowest branch, i.e. lower than 2 Ω.

5) The biggest current flows through the middle branch because that branch has the lowest resistance.

Everything Electrical in a Car is Connected in Parallel

Parallel connection is essential in a car to give these two features:

> 1) Everything can be turned on and off separately.
> 2) Everything always gets the full voltage from the battery.

The only slight effect is that when you turn lots of things on the lights may go dim because the battery can't provide full voltage under heavy load. This is normally a very slight effect. You can spot the same thing at home when you turn a kettle on, if you watch very carefully.

Ⓜ is the symbol for a motor.

In a parallel universe — my car would start...

A lot of fairy lights are actually done on a parallel circuit these days — they have an adapter that brings the voltage down, so the lights can still be diddy but it doesn't matter if one of them blows. Cunning.

Physics 2(ii) — Electricity and the Atom

Mains Electricity

Mains Supply is AC, Battery Supply is DC

1) The UK mains supply is approximately 230 volts.

2) It is an AC supply (alternating current), which means the current is constantly changing direction.

3) The frequency of the AC mains supply is 50 cycles per second or 50 Hz (50 hertz).

4) By contrast, cells and batteries supply direct current (DC). This just means that the current keeps flowing in the same direction.

AC Can Be Shown on an Oscilloscope Screen

1) A cathode ray oscilloscope (CRO) is basically a snazzy voltmeter.

2) If you plug an AC supply into an oscilloscope, you get a 'trace' on the screen that shows how the voltage of the supply changes with time. The trace goes up and down in a regular pattern — some of the time it's positive and some of the time it's negative.

3) The vertical height of the trace at any point shows the input voltage at that point.

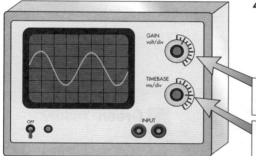

4) There are two dials on the front of the oscilloscope called the TIMEBASE and the GAIN. You can use these to set the scales of the horizontal and vertical axes of the display.

The GAIN dial controls how many volts each centimetre division represents on the vertical axis.

The TIMEBASE dial controls how many milliseconds (1 ms = 0.001 s) each division represents on the horizontal axis.

Learn How to Read an Oscilloscope Trace

DC supply

A DC source is always at the same voltage, so you get a straight line.

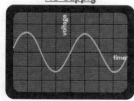

AC supply

An AC source gives a regularly repeating wave. From that, you can work out the period and the frequency of the supply.

You work out the frequency using:

$$\text{Frequency (Hz)} = \frac{1}{\text{Time period (s)}}$$

EXAMPLE: The trace below comes from an oscilloscope with the timebase set to 5 ms/div. Find: a) the time period, and b) the frequency of the AC supply.

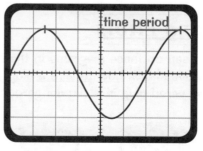

time period

ANSWER: a) To find the time period, measure the horizontal distance between two peaks. The time period of the signal is 6 divisions. Multiply by the timebase:
Time period = 5 ms × 6 = 0.03 s

b) Using the frequency formula on the left:
Frequency = 1/0.03 = 33 Hz

I wish my bank account had a gain dial...

Because mains power is AC, its current can be increased or decreased using a device called a transformer. The lower the current in power transmission lines, the less energy is wasted as heat.

Mains Electricity

Now then, did you know... electricity is dangerous. It can kill you. Well just watch out for it, that's all.

Hazards in the Home — Eliminate Them Before They Eliminate You

A likely exam question will show you a picture of domestic bliss but with various electrical hazards in the picture such as kids shoving their fingers into sockets and stuff like that, and they'll ask you to list all the hazards. This should be mostly common sense, but it won't half help if you already know some of the likely hazards, so learn these 9 examples:

1) Long cables.
2) Frayed cables.
3) Cables in contact with something hot or wet.
4) Water near sockets.
5) Shoving things into sockets.

6) Damaged plugs.
7) Too many plugs into one socket.
8) Lighting sockets without bulbs in.
9) Appliances without their covers on.

Plugs and Cables — Learn the Safety Features

Get the Wiring Right:

1) The right coloured wire to each pin, and firmly screwed in.
2) No bare wires showing inside the plug.
3) Cable grip tightly fastened over the cable outer layer.

Plug Features:

1) The metal parts are made of copper or brass because these are very good conductors.

2) The case, cable grip and cable insulation are all made of plastic or rubber because these are really good insulators and are flexible too.

3) This all keeps the electricity flowing where it should.

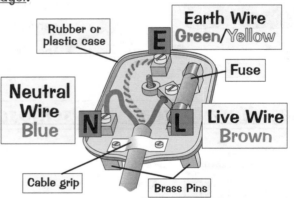

Rubber or plastic case

Earth Wire
Green/Yellow

Fuse

Neutral Wire
Blue

N

E

L

Live Wire
Brown

Cable grip

Brass Pins

Plug Wiring Errors

They're pretty keen on these diagrams in the exam so make sure you know them.
The diagram above shows how to wire a plug properly. Shown below are examples of how not to wire a plug. A badly wired plug is really dangerous so learn these diagrams.

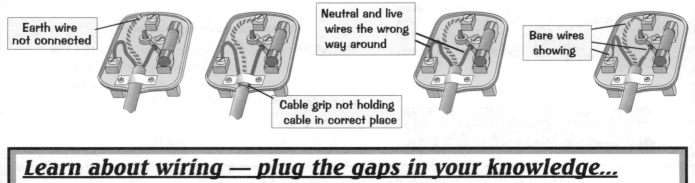

Earth wire not connected

Neutral and live wires the wrong way around

Cable grip not holding cable in correct place

Bare wires showing

Learn about wiring — plug the gaps in your knowledge...

Learning the stuff on this page is really important, and not just for your exam — one shock from the mains can kill you (okay, lecture over). Make sure you know how to wire a plug properly, and trickier, can spot when one's wired badly. Learnt it all? Cover the page and scribble it all down again.

Fuses and Earthing

Mains Cables Have Three Separate Wires

The brown LIVE WIRE in a mains supply alternates
between a HIGH +VE AND –VE VOLTAGE.
The blue NEUTRAL WIRE is always at 0V.
Electricity normally flows in and out through the
live and neutral wires only.
The green and yellow EARTH WIRE is just for
safety, and works together with a fuse to prevent
fire and shocks.

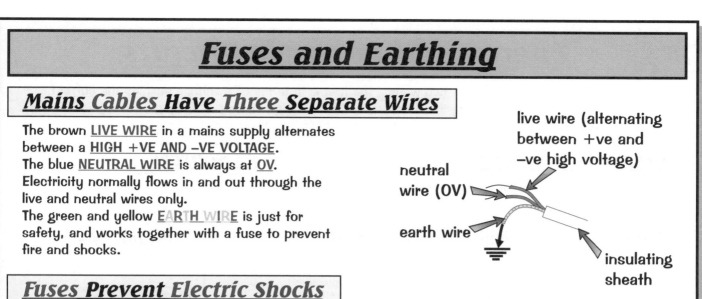

live wire (alternating
between +ve and
–ve high voltage)

neutral
wire (0V)

earth wire

insulating
sheath

Fuses Prevent Electric Shocks

1) To prevent surges of current in electrical circuits and danger of electric shocks, a fuse is
   normally placed in the circuit. [You can sometimes use a circuit breaker (a resettable fuse)
   instead, which works slightly differently.]

2) If the current in the circuit gets too big (bigger than the fuse rating), the fuse wire heats up
   and the fuse blows, breaking the circuit and preventing any electric shocks.

3) Fuses should be rated as near as possible but just higher than the normal operating current,
   (see next page).

4) The fuse should always be the same value as the manufacturer recommends.

Earthing Prevents Fires and Shocks

The EARTH WIRE and fuse work together like this:

1) The earth pin is connected to the case via the earth wire (the yellow and green wire).

2) If a fault develops in
which the live somehow
touches the metal case,
then because the case is
earthed, a big current
flows in through the live,
through the case and out
down the earth wire.

3) This surge in current
blows the fuse, which
cuts off the live supply.
This prevents electric
shocks from the case.

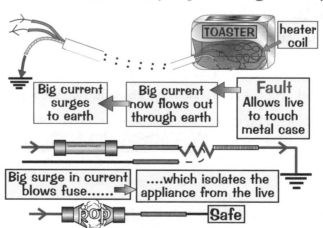

TOASTER heater
coil

Big current
surges
to earth

Big current
now flows out
through earth

Fault
Allows live
to touch
metal case

Big surge in current
blows fuse......

....which isolates the
appliance from the live

POP

Safe

All appliances with metal cases must be "earthed" to avoid the danger of electric shock.
"Earthing" just means attaching the metal case to the earth wire in the cable.

Why are earth wires green and yellow — when mud's brown...

Sometimes, you can use a Residual Current Circuit Breaker (RCCB) instead of a fuse and an earth wire.
RCCBs work a bit differently. Normally, exactly the same current flows through the live and neutral
wires, but if somebody touches the live wire a huge current flows through them to earth. That leaves
the neutral wire carrying less current than the live wire. The circuit breaker instantly detects this
difference in current and cuts off the power. RCCBs can be reset at the flick of a switch, so they're
much more convenient than fuses that have to be replaced each time they melt. Clever, eh?

Energy and Power in Circuits

You can look at <u>electrical circuits</u> in <u>two ways</u>. The first is in terms of a voltage <u>pushing the current</u> round and the resistances opposing the flow, as on P.85. The <u>other way</u> of looking at circuits is in terms of <u>energy transfer</u>. Learn them <u>both</u> and be ready to tackle questions about <u>either</u>.

Energy is Transferred from Cells and Other Sources

Anything which <u>supplies electricity</u> is also supplying <u>energy</u>.

So cells, batteries, generators, etc. all <u>transfer energy</u> to components in the circuit:

| <u>Motion</u>: motors | <u>Light</u>: light bulbs | <u>Heat</u>: Hairdriers/kettles | <u>Sound</u>: speakers |

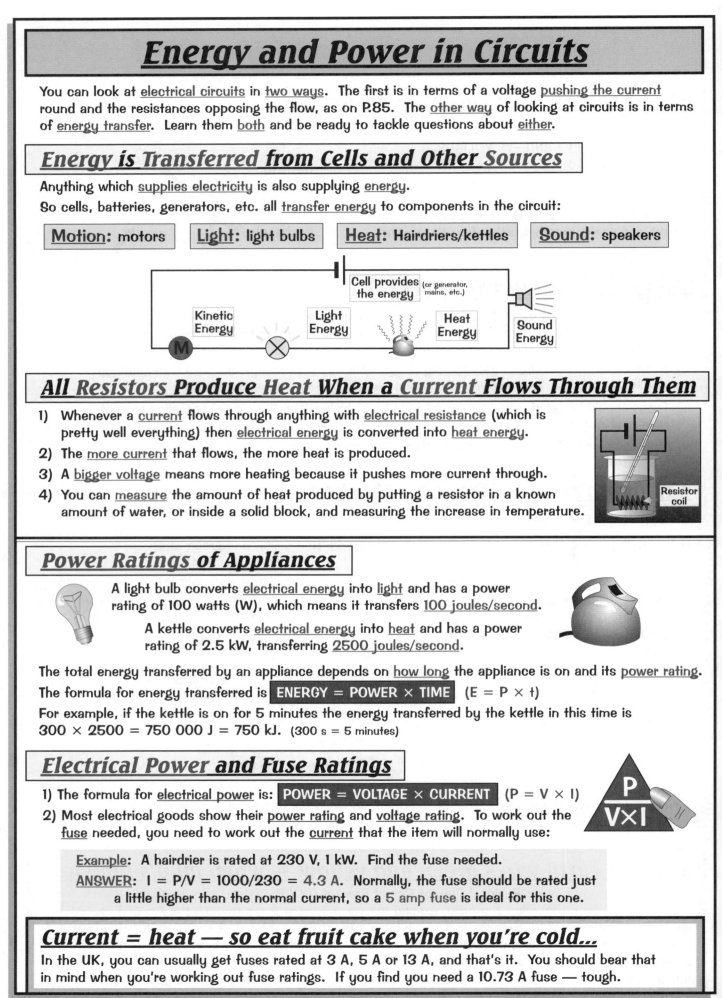

All Resistors Produce Heat When a Current Flows Through Them

1) Whenever a <u>current</u> flows through anything with <u>electrical resistance</u> (which is pretty well everything) then <u>electrical energy</u> is converted into <u>heat energy</u>.

2) The <u>more current</u> that flows, the more heat is produced.

3) A <u>bigger voltage</u> means more heating because it pushes more current through.

4) You can <u>measure</u> the amount of heat produced by putting a resistor in a known amount of water, or inside a solid block, and measuring the increase in temperature.

Power Ratings of Appliances

A light bulb converts <u>electrical energy</u> into <u>light</u> and has a power rating of 100 watts (W), which means it transfers <u>100 joules/second</u>.

A kettle converts <u>electrical energy</u> into <u>heat</u> and has a power rating of 2.5 kW, transferring <u>2500 joules/second</u>.

The total energy transferred by an appliance depends on <u>how long</u> the appliance is on and its <u>power rating</u>.

The formula for energy transferred is **ENERGY = POWER × TIME** (E = P × t)

For example, if the kettle is on for 5 minutes the energy transferred by the kettle in this time is 300 × 2500 = 750 000 J = 750 kJ. (300 s = 5 minutes)

Electrical Power and Fuse Ratings

1) The formula for <u>electrical power</u> is: **POWER = VOLTAGE × CURRENT** (P = V × I)

2) Most electrical goods show their <u>power rating</u> and <u>voltage rating</u>. To work out the <u>fuse</u> needed, you need to work out the <u>current</u> that the item will normally use:

> <u>Example</u>: A hairdrier is rated at 230 V, 1 kW. Find the fuse needed.
>
> <u>ANSWER</u>: I = P/V = 1000/230 = 4.3 A. Normally, the fuse should be rated just a little higher than the normal current, so a 5 amp fuse is ideal for this one.

Current = heat — so eat fruit cake when you're cold...

In the UK, you can usually get fuses rated at 3 A, 5 A or 13 A, and that's it. You should bear that in mind when you're working out fuse ratings. If you find you need a 10.73 A fuse — tough.

Physics 2(ii) — Electricity and the Atom

Charge, Voltage and Energy Change

Total Charge Through a Circuit Depends on Current and Time

1) Current is the <u>flow of electrical charge</u> (in coulombs, C) around a circuit.

2) When <u>current</u> (I) flows past a point in a circuit for a length of <u>time</u> (t) then the <u>charge</u> (Q) that has passed is given by this formula:

> **Total charge = Current × Time**

3) <u>More charge</u> passes around the circuit when a <u>bigger current</u> flows.

> <u>Example</u>: A battery charger passes a current of 2.5 A through a cell over a period of 4 hours. How much charge does the charger transfer to the cell altogether?
> <u>ANSWER</u>: Q = I × t = 2.5 × (4 × 60 × 60) = 36 000 C (36 kC).

The Voltage is the Energy Transferred per Charge Passed

1) When an electrical <u>charge</u> (Q) goes through a <u>change</u> in voltage (V), then <u>energy</u> (E) is <u>transferred</u>.

2) Energy is <u>supplied</u> to the charge at the <u>power source</u> to 'raise' it through a voltage.

3) The charge <u>gives up</u> this energy when it 'falls' through any <u>voltage drop</u> in <u>components</u> elsewhere in the circuit.

The formula is real simple:

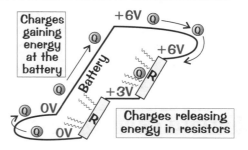

> **Energy transformed = Charge × Potential difference**

4) The <u>bigger</u> the <u>change</u> in voltage (or P.D.), the <u>more energy</u> is transferred for a <u>given amount of charge</u> passing through the circuit.

5) That means that a battery with a <u>bigger voltage</u> will supply <u>more energy</u> to the circuit for every <u>coulomb</u> of charge which flows round it, because the charge is raised up "<u>higher</u>" at the start (see above diagram) — and as the diagram shows, <u>more energy</u> will be <u>dissipated</u> in the circuit too.

> <u>Example</u>: The motor in an electric toothbrush is attached to a 3 V battery. If a current of 0.8 A flows through the motor for 3 minutes:
> a) Calculate the total charge passed.
> b) Calculate the energy transformed by the motor.
> c) Explain why the kinetic energy output of the motor will be less than your answer to b).
>
> <u>ANSWER</u>: a) Use the formula from the top of the page, Q = I × t = 0.8 × (3 × 60) = <u>144 C</u>
> b) Use E = Q × V = 144 × 3 = <u>432 J</u>
> c) The motor won't be 100% efficient. Some of the energy will be transformed into <u>sound and heat</u>.

Okay, so it's a dull page — but just two formulas...

Electricity's quite handy when you get down to it. Shame it's all so... well... dull. I'll tell you what, once you've learnt those two formulas and can do the examples, I reckon you've earned an extra special treat. Here's one I ate earlier, enjoy...

Atomic Structure

Ernest Rutherford didn't just pick the nuclear model of the atom out of thin air. It all started with a Greek fella called Democritus in the 5th Century BC. He thought that <u>all matter</u>, whatever it was, was made up of <u>identical</u> lumps called "atomos". And that's about as far as the theory got until the 1800s...

Rutherford Scattering **and the** Demise **of the** Plum Pudding

1) In 1804 <u>John Dalton</u> agreed with Democritus that matter was made up of <u>tiny spheres</u> ("atoms") that couldn't be broken up, but he reckoned that <u>each element</u> was made up of a <u>different type</u> of "atom".

2) Nearly 100 years later, J J Thomson discovered that <u>electrons</u> could be <u>removed</u> from atoms. So Dalton's theory wasn't quite right (atoms could be broken up). Thomson suggested that atoms were <u>spheres of positive charge</u> with tiny negative electrons <u>stuck in them</u> like plums in a <u>plum pudding</u>.

3) That "plum pudding" theory didn't last very long though. In 1909 <u>Ernest Rutherford</u> and his merry men tried firing <u>alpha particles</u> at <u>thin gold foil</u>. Most of them just went <u>straight through</u>, but the odd one came <u>straight back</u> at them, which was frankly a bit of a <u>shocker</u> for Ernie and his pals.

4) Being a pretty clued-up guy, Rutherford realised this meant that <u>most of the mass</u> of the atom was concentrated at the <u>centre</u> in a <u>tiny nucleus</u>, with a <u>positive charge</u>.

5) And that most of an atom is just <u>empty space</u>, which is also a bit of a <u>shocker</u> when you think about it.

Rutherford Came Up with the Nuclear Model of the Atom

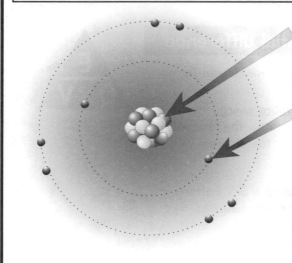

The <u>nucleus</u> is <u>tiny</u> but it makes up most of the <u>mass</u> of the atom. It contains <u>protons</u> (which are <u>positively charged</u>) and <u>neutrons</u> (which are <u>neutral</u>) — which gives it an overall positive charge.

The rest of the atom is mostly <u>empty space</u>. The <u>negative electrons</u> whizz round the outside of the nucleus really fast. They give the atom its <u>overall size</u>.

Learn the relative charges and masses of each particle:

PARTICLE	MASS	CHARGE
Proton	1	+1
Neutron	1	0
Electron	$\frac{1}{2000}$	-1

See Chemistry 2(i) for a few more details on this.

Isotopes are Different Forms of the Same Element

1) <u>Isotopes</u> are atoms with the <u>same</u> number of <u>protons</u> but a <u>different</u> number of <u>neutrons</u>.

2) Unstable isotopes are <u>radioactive</u>, which means they <u>decay</u> into <u>other elements</u> and <u>give out radiation</u>. This is where all <u>radioactivity</u> comes from — <u>unstable radioactive isotopes</u> undergoing <u>nuclear decay</u> and spitting out <u>high-energy particles</u>.

The Plum Pudding — by 1909 they'd had their fill of it...

The nuclear model is just <u>one way of thinking about</u> the atom. It works really well for explaining a lot of Chemistry, but it's certainly not the whole story. Other bits of science are explained using <u>different</u> models of the atom. The beauty of it though is that no one model is <u>more right</u> than the others.

Radioactive Decay Processes

When nuclei decay by alpha or beta emission, they change from one element to a different one.

Alpha Particles are Helium Nuclei

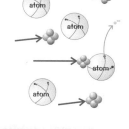

1) They are relatively <u>big</u> and <u>heavy</u> and <u>slow moving</u>.

2) They therefore <u>don't</u> penetrate very far into materials but are <u>stopped quickly</u>.

3) Because of their size they are <u>strongly</u> ionising, which just means they <u>bash into</u> a lot of atoms and <u>knock electrons off</u> them before they slow down, which creates lots of ions — hence the term "<u>ionising</u>".

Alpha Emission:

A typical <u>alpha emission</u>:

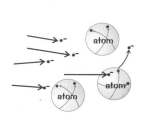

$$^{226}_{88}\text{Ra} \longrightarrow {}^{222}_{86}\text{Rn} \longrightarrow {}^{4}_{2}\text{He}$$

Unstable isotope · New isotope · Alpha particle

An <u>α-particle</u> is simply a <u>helium nucleus</u>, mass 4 and charge of +2, made up of 2 protons and 2 neutrons.

Beta Particles are Electrons

1) These are <u>in between</u> alpha and gamma in terms of their <u>properties</u>.

2) They move <u>quite</u> fast and they are <u>quite</u> small (they're electrons).

3) They <u>penetrate moderately</u> before colliding and are <u>moderately ionising</u> too.

4) For every $\beta-$ particle emitted, a <u>neutron</u> turns to a <u>proton</u> in the nucleus.

Beta Emission:

A typical <u>beta emission</u>:

$$^{14}_{6}\text{C} \longrightarrow {}^{14}_{7}\text{N} \longrightarrow {}^{0}_{-1}\text{e}$$

Unstable isotope · New isotope · Beta particle

A <u>β-particle</u> is simply an <u>electron</u>, with virtually no mass and a charge of –1. <u>Every time</u> a beta particle is emitted from a nucleus, a <u>neutron</u> in the nucleus is <u>converted</u> to a <u>proton</u>.

Gamma Rays are Very Short Wavelength EM Waves

1) They are the <u>opposite</u> of alpha particles in a way.

2) They <u>penetrate a long way</u> into materials without being stopped.

3) This means they are <u>weakly</u> ionising because they tend to <u>pass through</u> rather than colliding with atoms. Eventually they <u>hit something</u> and do <u>damage</u>.

Gamma Emission:

A typical combined α and γ emission:

$$^{238}_{92}\text{U} \longrightarrow {}^{234}_{90}\text{Th} \longrightarrow {}^{4}_{2}\text{He}$$
$$^{0}_{0}\gamma$$

Unstable isotope · New isotope · Gamma ray

A <u>γ-ray</u> is a <u>photon</u> with no mass and no charge.

After an <u>alpha or beta emission</u> the nucleus sometimes has <u>extra energy to get rid of</u>. It does this by emitting a <u>gamma ray</u>. Gamma emission <u>never changes</u> the <u>proton or mass numbers</u> of the nucleus.

I once beta particle — it cried for ages...

Learn all the details about the three different types of radiation — alpha, beta and gamma. When a nucleus decays by <u>alpha</u> emission, its <u>atomic number</u> goes down by <u>two</u> and its <u>mass number</u> goes down by <u>four</u>. <u>Beta</u> emission increases the atomic number by <u>one</u> (the mass number <u>doesn't change</u>).

Background Radiation

Background Radiation Comes from Many Sources

Background radiation we receive comes from:

1) Radioactivity of naturally occurring unstable isotopes which are all around us — in the air, in food, in building materials and in the rocks under our feet.

2) Radiation from space, which is known as cosmic rays. These come mostly from the Sun.

3) Radiation due to human activity, e.g. fallout from nuclear explosions, or dumped nuclear waste. But this represents a tiny proportion of the total background radiation.

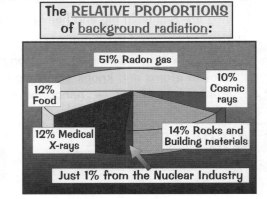

The RELATIVE PROPORTIONS of background radiation:

51% Radon gas
10% Cosmic rays
12% Food
12% Medical X-rays
14% Rocks and Building materials
Just 1% from the Nuclear Industry

The Level of Background Radiation Changes Depending on Where You Are

1) At high altitudes (e.g. in jet planes) the background radiation increases because of more exposure to cosmic rays. That means commercial pilots have an increased risk of getting some types of cancer.

2) Underground in mines, etc. it increases because of the rocks all around.

3) Certain underground rocks (e.g. granite) can cause higher levels at the surface, especially if they release radioactive radon gas, which tends to get trapped inside people's houses.

Radon Gas is the Subject of Scientific Debate

1) The radon concentration in people's houses varies widely across the UK, depending on what type of rock the house is built on.

2) Studies have shown that exposure to high doses of radon gas can cause lung cancer — and the greater the radon concentration, the higher the risk.

3) The scientific community is a bit divided on the effects of lower doses, and there's still a lot of debate over what the highest safe(ish) concentration is.

4) Evidence suggests that the risk of developing lung cancer from radon is much greater for smokers compared to non-smokers.

5) Some medical professionals reckon that about 1 in 20 deaths from lung cancer (about 2000 per year) are caused by radon exposure.

6) New houses in areas where high levels of radon gas might occur must be designed with good ventilation systems. These reduce the concentration of radon in the living space.

7) In existing houses, the Government recommends that ventilation systems are put in wherever the radiation due to radon is above a certain level.

Coloured bits indicate more radiation from rocks

Background radiation — it's like nasty wallpaper...

Did you know that background radiation was first discovered accidentally. Scientists were trying to work out which materials were radioactive, and couldn't understand why their reader still showed radioactivity when there was no material being tested. They realised it must be natural background radiation.

Nuclear Fission and Fusion

Nuclear Fission — the Splitting Up of Big Atomic Nuclei

Nuclear power stations and nuclear submarines are both powered by nuclear reactors.

In a nuclear reactor, a controlled chain reaction takes place in which atomic nuclei split up and release energy in the form of heat. This heat is then simply used to heat water to drive a steam turbine. So nuclear reactors are really just glorified steam engines! The "fuel" that's split is usually either uranium-235 or plutonium-239 (or both).

Steam generator
Control rods
Coolant pump
Steam to turbine
Return water
Pressurised Coolant
Uranium fuel rods

The Chain Reactions:

1) If a slow-moving neutron gets absorbed by a uranium or plutonium nucleus, the nucleus can split.

2) Each time a uranium or plutonium nucleus splits up, it spits out two or three neutrons, one of which might hit another nucleus, causing it to split also, and thus keeping the chain reaction going.

3) When a large atom splits in two it will form two new lighter elements. These new nuclei are usually radioactive because they have the "wrong" number of neutrons in them. This is the big problem with nuclear power — it produces huge amounts of radioactive material which is very difficult and expensive to dispose of safely.

4) Each nucleus splitting (called a fission) gives out a lot of energy — a lot more energy than you get with a chemical bond between two atoms. Make sure you remember that. Nuclear processes release much more energy than chemical processes do. That's why nuclear bombs are so much more powerful than ordinary bombs (which rely on chemical reactions).

Nuclear Fusion — the Joining of Small Atomic Nuclei

1) Two light nuclei (e.g. hydrogen) can combine to create a larger nucleus — this is called nuclear fusion.

2) Fusion releases a lot of energy — all the energy released in stars comes from fusion. So people are trying to develop fusion reactors to make electricity.

3) Fusion doesn't leave behind a lot of radioactive waste and there's plenty of hydrogen knocking about to use as fuel.

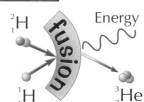

4) The big problem is that fusion can only happen at really high temperatures — about 10 000 000 °C.

5) No material can stand that kind of temperature without being vaporised, so fusion reactors are really hard to build. You have to contain the hot hydrogen in a magnetic field instead of a physical container.

6) There are a few experimental reactors around, but none of them are generating electricity yet. At the moment it takes more power to get up to temperature than the reactor can produce.

Ten million degrees — that's hot...

At about the same time as research started on fusion reactors, physicists were working on a fusion bomb. These "hydrogen bombs" are incredibly powerful — they can release a few thousand times more energy than the nuclear bombs that destroyed Hiroshima and Nagasaki at the end of World War II.

Revision Summary for Physics 2(ii)

There's some pretty heavy physics in this section. But just take it one page at a time and it's not so bad. When you think you know it all, try these questions and see how you're getting on. If there are any you can't do, look back at the right bit of the section, learn it, then come back here and try again.

1) What causes the build-up of static electricity? Which particles move when static builds up?

2) Give two examples of how static electricity can be helpful. Write all the details.

3) Give two examples each of static electricity being: a) a nuisance, b) dangerous.

4) Explain what current, voltage and resistance are in an electric circuit.

5) Sketch typical voltage-current graphs for: a) a resistor, b) a filament lamp, c) a diode. Explain the shape of each graph.

6)* Calculate the resistance of a wire if the voltage across it is 12 V and the current through it is 2.5 A.

7) Describe how the resistance of an LDR varies with light intensity. Give an application of an LDR.

8)* Find each unknown voltage, current or resistance in this circuit.

9) Why are parallel circuits often more useful than series ones?

10)* An oscilloscope is plugged into the mains (50 Hz). Sketch what you would expect to see on the screen if the timebase is set to 2 ms/div.

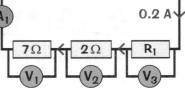

11) Sketch a properly wired three-pin plug.

12) Explain fully how a fuse and earth wire work together.

13)* Find the appropriate fuse (3 A, 5 A or 13 A) for these appliances:
a) a toaster rated at 230 V, 1100 W   b) an electric heater rated at 230 V, 2000 W

14)* Calculate the energy transformed by a torch using a 6 V battery when 530 C of charge pass through.

15) Describe the "plum pudding" model of the atom.

16) Describe Rutherford's scattering experiment with a diagram. What were the results of the experiment, and what did Rutherford conclude from it?

17) Draw a table stating the relative mass and charge of the three basic subatomic particles.

18)* Write down the number of protons, neutrons and electrons in an atom of $^{230}_{90}$Th.

19) Describe in detail the nature and properties of the three types of radiation: α, β and γ.

20)* Write down the nuclear equation for the alpha decay of: a) $^{234}_{92}$U, b) $^{230}_{90}$Th and c) $^{226}_{88}$Ra.

21)* Write down the nuclear equation for the combined beta and gamma decay of:
a) $^{234}_{90}$Th, b) $^{234}_{91}$Pa and c) $^{14}_{6}$C.

22) List three places where the level of background radiation is increased and explain why.

23) Draw a diagram to illustrate the fission of uranium-235 or plutonium-239 and explain how the chain reaction works.

24) What is nuclear fusion? Why is it difficult to construct a working fusion reactor?

* Answers at the bottom of the page

Physics 2(ii) — Electricity and the Atom

Thinking in Exams

In the old days, it was enough to learn a whole bunch of <u>facts</u> while you were revising and just spew them onto the paper come exam day. If you knew the facts, you had a good chance of doing well, even if you didn't really <u>understand</u> what any of those facts actually meant. But those days are over. Rats.

Remember — You Might Have to Think During the Exam

1) Nowadays, the examiners want you to be able to <u>apply</u> your scientific knowledge and <u>understand articles</u> written about science. Eeek.

2) The trick is <u>not</u> to <u>panic</u>. They're <u>not</u> expecting you to show Einstein-like levels of scientific insight (not usually, anyway).

3) They're just expecting you to use the science you <u>know</u> in both <u>familiar</u> and <u>unfamiliar settings</u>. And sometimes they'll give you some <u>extra info</u> too that you should use in your answer.

So to give you an idea of what to expect come exam-time, use the new <u>CGP Exam Simulator</u> (below). Read the article, and have a go at the questions. It's <u>guaranteed</u> to be just as much fun as the real thing.

Underlining or making notes of the main bits as you read is a good idea.

1. Stopping distance divided into distance for thinking and braking

2. Distance varies
→ speed, driver, car

3. Car needs to have good grip on road...
to improve stopping distances...:

    4. Grip on road
      → tyres
      → weather conditions
      → road surface

The distance it takes for a car to stop once the driver has seen a hazard is divided between the <u>thinking distance</u> and the <u>braking distance</u>. These two distances are affected by many factors, such as <u>how fast the car is going</u> and the <u>condition of the driver and the car.</u>

It's important for the car to have a <u>good grip</u> on the road. The better the grip <u>the sooner the car will be able to stop</u>. Tyres should have a <u>minimum tread depth</u> of 1.6 mm to have enough grip in wet conditions. Without any tread the tyre will just ride on the layer of water and skid really easily. Drivers should take extra care when it's <u>wet or icy as</u> the road's slippier than when it's dry. Even the road surface <u>can make</u> a difference — gravel, leaves and muck can all cause the car to slip.

Mark Smith, a director of GoodTyres plc, said: "We recommend regularly replacing your tyres for the best control of your car and better road safety."

<u>Questions</u>:

1. Why should a car owner check the tread depth of their car's tyres?

2. In rainy conditions, why should people leave a greater distance between their car and the car in front?

3. Why might some people suspect Mark Smith of being biased?

Clues — don't read unless you need a bit of a hand...
1. Think about what would happen if the tyres didn't have any tread.
2. What happens to the road surface in wet conditions? What effect will this have on the stopping distance?
3. What's his job?

Answers

1) The tyre tread is important for preventing skidding in wet conditions — so its depth should be checked regularly.

2) In wet conditions there is more water on the road, which means the car has less grip. This increases the braking distance and so the overall stopping distance.

3) He's a director of a firm that probably makes tyres — so he'll want to make them sound as important as possible.

Don't skim read — you might something...

It's so easy to skim read an article given to you in an exam. But don't. Read it really well, underlining or making notes as you go. It's well worth spending some time making sure you understand the article and what the questions are asking for before scribbling down your answers.

Answering Experiment Questions (i)

Science is all (well... a lot) about doing experiments carefully, and interpreting results.
And so that's what they're going to test you on when you do your exam. Among other things.

Read the Question Carefully

Expect at least some questions to describe experiments — a bit like the one below.

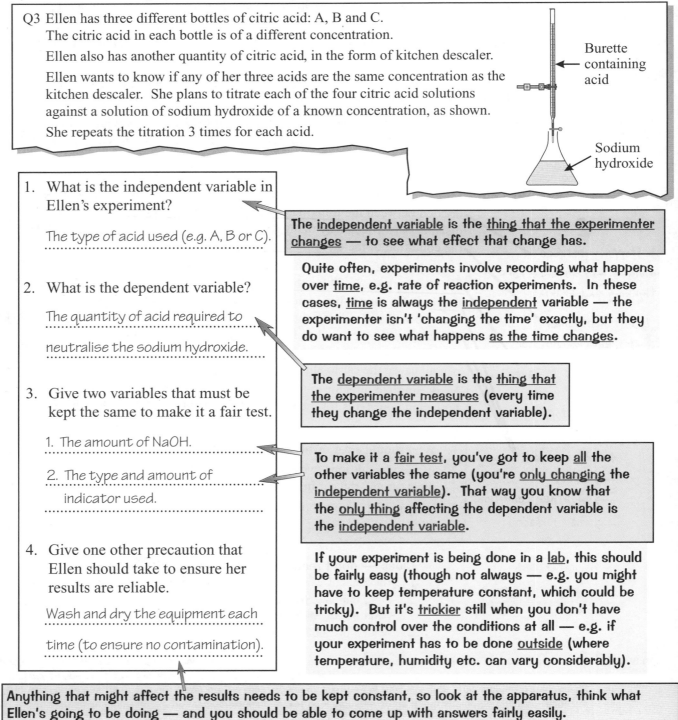

Q3 Ellen has three different bottles of citric acid: A, B and C.
The citric acid in each bottle is of a different concentration.

Ellen also has another quantity of citric acid, in the form of kitchen descaler.

Ellen wants to know if any of her three acids are the same concentration as the kitchen descaler. She plans to titrate each of the four citric acid solutions against a solution of sodium hydroxide of a known concentration, as shown.

She repeats the titration 3 times for each acid.

Burette containing acid

Sodium hydroxide

1. What is the independent variable in Ellen's experiment?

    The type of acid used (e.g. A, B or C).

The **independent variable** is the **thing that the experimenter changes** — to see what effect that change has.

Quite often, experiments involve recording what happens over **time**, e.g. rate of reaction experiments. In these cases, **time** is always the **independent** variable — the experimenter isn't 'changing the time' exactly, but they do want to see what happens **as the time changes**.

2. What is the dependent variable?

    The quantity of acid required to neutralise the sodium hydroxide.

The **dependent variable** is the **thing that the experimenter measures** (every time they change the independent variable).

3. Give two variables that must be kept the same to make it a fair test.

    1. The amount of NaOH.

    2. The type and amount of indicator used.

To make it a **fair test**, you've got to keep **all** the other variables the same (you're **only changing** the **independent variable**). That way you know that the **only thing** affecting the dependent variable is the **independent variable**.

4. Give one other precaution that Ellen should take to ensure her results are reliable.

    Wash and dry the equipment each time (to ensure no contamination).

If your experiment is being done in a **lab**, this should be fairly easy (though not always — e.g. you might have to keep temperature constant, which could be tricky). But it's **trickier** still when you don't have much control over the conditions at all — e.g. if your experiment has to be done **outside** (where temperature, humidity etc. can vary considerably).

Anything that might affect the results needs to be kept constant, so look at the apparatus, think what Ellen's going to be doing — and you should be able to come up with answers fairly easily.

If the equipment isn't <u>clean</u>, that will definitely affect the results. And if the flask's not <u>dry</u>, the extra water would dilute the sodium hydroxide slightly (which would affect the results). A change in temperature could also be a problem (though probably a small one) — things expand as they get hotter, so Ellen could get a false reading from the burette if the temperature in the lab changes drastically between tests.

Answering Experiment Questions (ii)

5. Why did Ellen repeat the titration 3 times for each acid?

   To check for anomalous results and make

   the results more reliable.

Sometimes you get <u>unusual results</u> — <u>repeating</u> an experiment gives you a better idea what the <u>correct result</u> should be.

6. The table below shows the amount of acid required in each titration.

	1st result (cm³)	2nd result (cm³)	3rd result (cm³)	Mean (cm³)
Kitchen descaler	24.1	23.9	23.7	
Acid A	23.9	23.5	24.0	23.8
Acid B	33.3	33.7	(38.6)	33.5
Acid C	23.7	23.9	24.1	23.9

When an experiment is <u>repeated</u>, the results will usually be <u>slightly different</u> each time.

To get a single <u>representative</u> value, you'd usually find the <u>mean</u> (average) of all the results.

The more times the experiment is <u>repeated</u> the <u>more reliable</u> this average will be.

To find the mean:

ADD TOGETHER all the data values and DIVIDE by the total number of values in the sample.

The <u>range</u> is how <u>spread out</u> the data is.

You just work out the <u>difference</u> between the <u>highest</u> and <u>lowest</u> numbers.

   a) Calculate the mean amount of kitchen descaler required to neutralise the NaOH.

   Mean = (24.1 + 23.9 + 23.7) ÷ 3 = 23.9 cm³

   b) What is the range of the quantities of kitchen descaler required?

   24.1 − 23.7 = 0.4 cm³

If one result doesn't seem to fit in — it's <u>wildly out</u> compared to all the others — then it's called an <u>anomalous</u> result. You should usually <u>ignore</u> an anomalous result (or even better — investigate it and try to work out what happened). Here, it's been <u>ignored</u> when the mean was worked out.

This one's a <u>random error</u> — one that only happens occasionally.

7. One of the results on the table is anomalous. Circle the result and suggest why it may have occurred.

   The reading may not have been taken

   correctly, or the wrong quantity of

   NaOH may have been used.

If you make the same mistake every time, it's a <u>systematic error</u>.

For example, if you measured the volume of a liquid using the <u>top</u> of the meniscus rather than the <u>bottom</u>, all your readings would be a little on the large side.

This reading should be 24.5 cm³ → 25.0 / 24.0 / 23.0

8. Using these results, which acid can you conclude is <u>not</u> the same concentration as the kitchen descaler?

   Acid B

You have to be careful here — both Acids A and C could be the same concentration, since all experiments have a "margin of error" — meaning results are never absolutely spot on.

So you can say that Acid B has a different concentration — but Acids A and C could be the same.

We all make mistakes...

No scientist does an experiment just once — unless they like people to point and laugh when the result turns out to be <u>wrong</u>. It's like weightlifting — the more times you repeat an experiment, the better the results will be (unless you're making <u>systematic</u> errors — you'd just have <u>lots</u> of <u>wrong results</u> then).

Exam Skills

Answering Experiment Questions (iii)

Use Sensible Measurements for Your Variables

Pu-lin did an experiment to see how the mass of a potato changed depending on the sugar solution it was in.

She started off by making potato tubes 5 cm in length, 1 cm in diameter and 2.0 g in mass. She then filled a beaker with 500 ml of pure water and placed a potato tube in it for 30 minutes. She repeated the experiment with different amounts of sugar dissolved in the water. For each potato tube, she measured the new mass. She did the experiment using Charlotte, Desiree, King Edward and Maris Piper potatoes.

Before she started, she did a trial run, which showed that most of the potato tubes shrunk to a minimum of 1 g (in a really strong sugar solution) or grew to a maximum of 3 g (in pure water).

1. What kind of variable was the list of potatoes?

   A A continuous variable ☐

   B A categoric variable ✓

   C An ordered variable ☐

   D A discrete variable ☐

2. Pu-lin should add sugar in intervals of...

   A a pinch ☐

   B a teaspoon ✓

   C a cupful ☐

   D a bucketful ☐

3. The balance used to find the mass of the potato should be capable of measuring...

   A to the nearest 0.01 gram ✓

   B to the nearest 0.1 gram ☐

   C to the nearest gram ☐

   D to the nearest 10 grams ☐

Categoric variables are variables that can't be related to size or quantity — they're <u>types</u> of things. E.g. <u>names of potatoes</u> or <u>types of fertiliser</u>.

<u>Continuous data</u> is <u>numerical data</u> that can have <u>any value</u> within a range — e.g. length, volume, temperature and time. Note: You <u>can't</u> measure the <u>exact value</u> of continuous data. Say you measure a height as 5.6 cm to the nearest mm. It's not <u>exact</u> — you get a more precise value if you measure to the nearest 0.1 mm or 0.01 mm, etc.

<u>Ordered variables</u> are things like <u>small, medium and large lumps</u>, or <u>warm, very warm and hot</u>.

<u>Discrete data</u> is the type that can be counted in chunks, where there's no in-between value. E.g. <u>number of people</u> is discrete, not continuous, because you can't have half a person.

It's important to use <u>sensible values</u> for variables.
It's no good using <u>loads</u> of sugar or <u>really weedy amounts</u> like a pinch at a time cos you'd be there <u>forever</u> and the results wouldn't show any <u>significant difference</u>. (You'd get different amounts of sugar in each pinch anyway.)

A balance measuring only to the nearest gram, or bigger, would <u>not</u> be <u>sensitive enough</u> — the changes in mass are likely to be quite small, so you'd need to measure to the <u>nearest 0.01 gram</u> to get the <u>most precise</u> results.

The <u>sensitivity</u> of an instrument is the <u>smallest change</u> it can detect, e.g. some balances measure to the nearest <u>gram</u>, but really sensitive ones measure to the nearest <u>hundredth of a gram</u>.
For measuring <u>tiny changes</u> — like from 2.00 g to 1.92 g — the more sensitive balance is needed.

You also have to think about the <u>precision</u> and <u>accuracy</u> of your results.

Precise results are ones <u>taken with sensitive instruments</u>, e.g. volume measured with a burette will be <u>more</u> precise than volume measured with a 100 ml beaker. Really accurate results are those that are <u>really close</u> to the <u>true answer</u>. It's possible for results to be precise but not very accurate, e.g. a fancy piece of lab equipment might give results that are precise, but if it's not calibrated properly those results won't be accurate.

I take my tea milky with two bucketfuls of sugar... mmm...

Accuracy, precision and sensitivity are difficult things to get your head around — a sensitive piece of equipment is likely to give precise results (but not necessarily very accurate results). If the equipment is used properly and calibrated well then the results are more likely to be accurate...

Answering Experiment Questions (iv)

Once you've collected all your data together, you need to analyse it to find any relationships between the variables. The easiest way to do this is to draw a graph, then describe what you see...

Graphs Are Used to Show Relationships

These are the results Pu-lin obtained with the King Edward potato.

Number of teaspoons of sugar	0	2	4	6	8	10	12	14	16	18	20
Mass of potato tube (g)	2.50	2.40	2.23	2.10	2.02	1.76	1.66	1.25	1.47	1.3	1.15

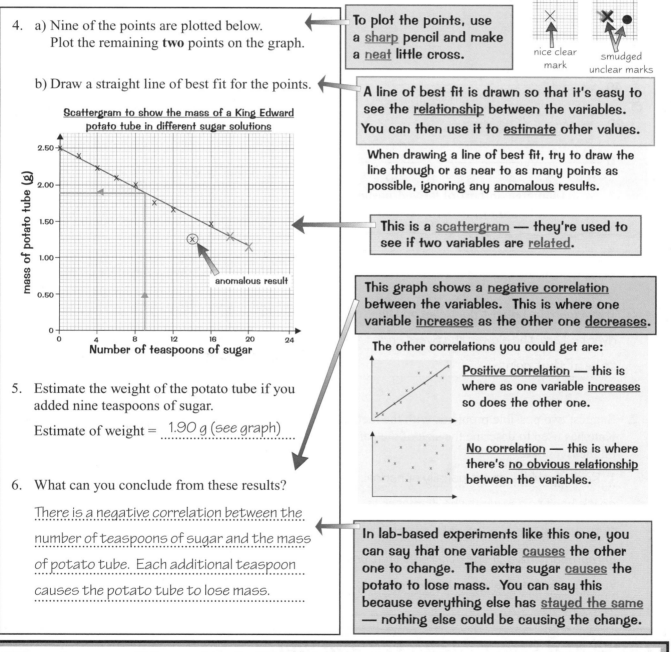

4. a) Nine of the points are plotted below.
       Plot the remaining **two** points on the graph.

    b) Draw a straight line of best fit for the points.

To plot the points, use a <u>sharp</u> pencil and make a <u>neat</u> little cross.

nice clear mark

smudged unclear marks

A line of best fit is drawn so that it's easy to see the <u>relationship</u> between the variables. You can then use it to <u>estimate</u> other values.

When drawing a line of best fit, try to draw the line through or as near to as many points as possible, ignoring any <u>anomalous</u> results.

Scattergram to show the mass of a King Edward potato tube in different sugar solutions

mass of potato tube (g)

Number of teaspoons of sugar

anomalous result

This is a <u>scattergram</u> — they're used to see if two variables are <u>related</u>.

This graph shows a <u>negative correlation</u> between the variables. This is where one variable <u>increases</u> as the other one <u>decreases</u>.

The other correlations you could get are:

<u>Positive correlation</u> — this is where as one variable <u>increases</u> so does the other one.

<u>No correlation</u> — this is where there's <u>no obvious relationship</u> between the variables.

5. Estimate the weight of the potato tube if you added nine teaspoons of sugar.

    Estimate of weight = <u>1.90 g (see graph)</u>

6. What can you conclude from these results?

    <u>There is a negative correlation between the</u>
    <u>number of teaspoons of sugar and the mass</u>
    <u>of potato tube. Each additional teaspoon</u>
    <u>causes the potato tube to lose mass.</u>

In lab-based experiments like this one, you can say that one variable <u>causes</u> the other one to change. The extra sugar <u>causes</u> the potato to lose mass. You can say this because everything else has <u>stayed the same</u> — nothing else could be causing the change.

There's a positive correlation between revising and good marks...

...really, it's true. Other ways to improve your marks are to practise plotting graphs, and learning how to read them properly — make sure you're reading off the right axis for a start, and don't worry about drawing lines on the graph if it helps you to read it. Always double-check your answer... just in case.

Exam Skills

Answering Experiment Questions (v)

A lot of Physics experiments can be done in a <u>nice controlled way</u> in a laboratory. But not all of them. And once you get <u>out of the lab</u> and into the <u>real world</u>, it gets much <u>harder to control</u> all the <u>variables</u>.

Relationships Do NOT Always Tell Us the Cause

On holiday in Scotland, Kate found that mountain streams can be difficult to wade across — the streams flow quite slowly, but there are often many large rocks on the stream bed, which make it difficult to balance.

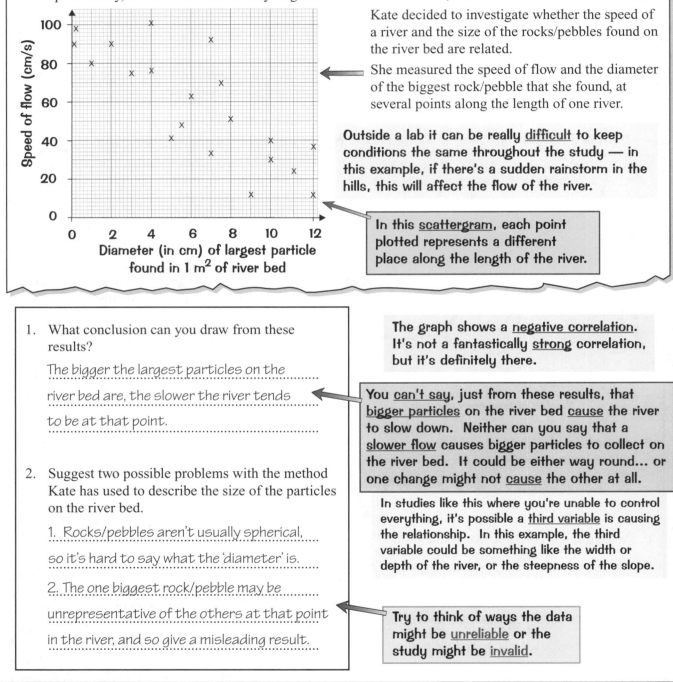

Kate decided to investigate whether the speed of a river and the size of the rocks/pebbles found on the river bed are related.

She measured the speed of flow and the diameter of the biggest rock/pebble that she found, at several points along the length of one river.

Outside a lab it can be really <u>difficult</u> to keep conditions the same throughout the study — in this example, if there's a sudden rainstorm in the hills, this will affect the flow of the river.

In this <u>scattergram</u>, each point plotted represents a different place along the length of the river.

1. What conclusion can you draw from these results?

   The bigger the largest particles on the river bed are, the slower the river tends to be at that point.

2. Suggest two possible problems with the method Kate has used to describe the size of the particles on the river bed.

   1. Rocks/pebbles aren't usually spherical, so it's hard to say what the 'diameter' is.

   2. The one biggest rock/pebble may be unrepresentative of the others at that point in the river, and so give a misleading result.

The graph shows a <u>negative correlation</u>. It's not a fantastically <u>strong</u> correlation, but it's definitely there.

You <u>can't say</u>, just from these results, that <u>bigger particles</u> on the river bed <u>cause</u> the river to slow down. Neither can you say that a <u>slower flow</u> causes bigger particles to collect on the river bed. It could be either way round... or one change might not <u>cause</u> the other at all.

In studies like this where you're unable to control everything, it's possible a <u>third variable</u> is causing the relationship. In this example, the third variable could be something like the width or depth of the river, or the steepness of the slope.

Try to think of ways the data might be <u>unreliable</u> or the study might be <u>invalid</u>.

Cause and effect — chicken and egg...

It's really hard to prove <u>causation</u>. Think about it — it sounds sensible that a rockier river bed provides more <u>resistance</u> to the flow of water, slowing it down. It sounds just as sensible that as the river slows down, there isn't enough 'oomph' to keep the big rocks bowling along. Which came first? Even if you reckon you <u>know</u> the answer — if your results <u>don't prove it</u>, you <u>can't</u> put it in your conclusion.

Index

A

acceleration 72, 73, 75, 77
accuracy 103, 105
acetate rod 83
acids 64
   reacting with metals 65
   reacting with metal oxides/
   hydroxides 66
activation energy 58, 59
aerobic respiration 17
air bags 81
air resistance 77
alginate bead 20
alkali metals 38
alkalis 64, 66
allele 32
allergies 20
alpha particles 96, 97
alternating current (AC) 91
amino acids 9, 10, 16, 17, 26
ammeter 85, 87
ammonia 42, 62, 66, 69
amps 85
amylase 18, 19
anhydrous copper(II) sulfate 60
anions 41
anode 68, 70
anomolous 103, 105
aquaplaning 78
A_r 47
artificial light 8
asexual reproduction 27, 28
atom economy 51
atomic number 36, 38
atomic structure 96
atoms 36, 37, 96
attraction (electrostatic) 83

B

baby foods 20
background radiation 98
bacteria 19, 27
balanced forces 75
balancing equations 46
Banting, Frederick 24
base 64
batteries 85, 87, 90
battery supply 91
Best, Charles 24
beta particles 97
big spark 84
bile 18, 19
biodegradable 20
biological detergents 20
blindness 24
blood sugar level 21, 23, 24
blood vessels 21
body temperature 21
boiling point 40, 43
bone marrow 29
BOOM! 84
braking distance 78
brine 69
buckminster fullerene 45
burglar detectors 87
burning fuels 60
by-products 51

C

cables, electrical 92
calculating masses 49
calculating resistance 86
car crash 81
carbohydrase 20
carbohydrate 23
carbon cycle 14
carbon dating 37
carbon dioxide 6, 7, 8, 14, 17, 21
carbonates 67
carrier 33
cars, wiring 90

catalysts 16, 54, 57-59, 61, 62
categoric variables 104
cathode 68, 70
cathode ray oscilloscope 91
cation 41
cause 105, 106
cell membrane 2, 4, 33
cell wall 2, 9
cells 2, 3, 87, 88, 94
cellulose 9
chain reaction 99
change in mass 55
charge 83, 84, 95
chemical bond 37
chloride (Cl⁻) ions 68
chloride salts 65-67
chlorine 42, 69
chlorophyll 2, 6, 10
chloroplasts 2
christmas fairy lights 90
chromosomes 26-28, 30
circuit breaker 93
circuits 85-90, 94, 95
clones 26
closed system 61
clothing crackles 84
cloudy precipitate 57
collision theory 58
combustion 60
components (electrical) 85
compounds 37
compromise 62
concentration 54, 56-58
concentration gradient 4
conduct electricity 40, 43
conductors 44, 83, 84
conservation of momentum 81
constant velocity 75
continuous variables 104
copper 65, 70
correlation 105, 106
cosmic rays 98
coulomb 95
covalent bonding 42, 43
cracking hydrocarbons 59
crazy acrobatics 32
CRO (oscilloscope) 91
crumple zones 81
current 85, 95
current, electrical 85, 87-89, 93, 94
current surges 93
cyanide 17
cystic fibrosis 33
cytoplasm 2, 27

D

Dalton, John 96
decay 12
deceleration 73, 75, 76
decomposition 57
Democritus 96
dependent variable 56, 102
detritus feeders 14
diabetes 23, 24, 29
diamond 43
differentiation 29
diffusion 4, 5
digestion 18
digestive system 19
diode 86, 87
direct current (DC) 91
discrete variables 104
disinfectants 69
displacement 67
distance travelled 73
distance-time graphs 73
DNA 26, 27
DNA fingerprinting 26
domestic electricity 92
dominant alleles 31-34
drinking water 67

driving force 85
dust removal 84
dynamic equilibrium 61, 62

E

earth wire 92, 93
earthing 84, 93
effort 79
egg cells 3, 28
elastic potential energy 80
electricity 68, 83-95
   electric shocks 93
   electrical charge 95
   electrical energy 94
   electrical hazards 92
   electrical power 94
electrolysis 68, 69, 70
electrolyte 68
electron configuration 39, 41
electron shell 38, 39, 41
electronic structure 41
electrons 36, 38, 68, 83-85, 96
element 37, 38, 96
EM waves/radiation 97
embryonic screening 33
empirical formula 48
emulsification 18
endothermic reaction 60, 61
energy 12, 13, 17, 58, 60,
   79, 94, 95
   energy of waves 94
   energy transferred 79
enzymes 8, 16-20
   digestive 18, 19
equal and opposite force 76
equilibrium 52, 61, 62
error 103
evaporate 67
exercise 23
exothermic reaction 60, 61
explosion 54
extra special treat 95

F

$F = ma$ 76
faeces 19
fair test 102
fallout, nuclear 98
faster collisions 58
fats 18
fatty acids 18
faulty cells 29
female characteristics 30
fertilisation 28
fertilisers 8, 62, 66
fertility clinics 29
filament lamp 86, 87
filtration 52, 67
fires 93
fission, nuclear 99
food chains 12
food production 13
force 74, 75, 79, 81
force diagram 76
force of gravity 74
forces 75, 76
forensic science 26
free electrons 44
free ions 68
free-falling objects 77
frequency 91
friction 77, 79, 83
fructose syrup 20
fruit 9
fruit cake 94
fuel filling nightmare 84
full outer shell 42
fullerenes 45
fuse 92-94
fuse rating 87, 94
fusion 99

G

gain (oscilloscope) 91
gall bladder 18, 19
gametes 28, 30
gamma rays 97
gangrene 24
gas syringe 55, 56, 57
genes 26, 31-33
genetic diagrams 30, 32, 34
genetic disorders 33
genetic engineering 24
genetic experiments 31
genetic fingerprinting 26
giant covalent structure 43, 44
giant ionic structure 40
glorified steam engines 99
glucagon 23
glucose 6, 9, 17, 23
glucose syrup 20
glucose-monitoring device 23
glycerol 18
gradient 73, 86
grain chutes 84
graphite 43
graphs 54, 56, 105, 106
gravitational potential energy (PE)
   80
gravity 74, 77
greenhouse 8
grip 78
groups 38
guard cells 3
gullet 19

H

H⁺ ions 64
Haber process 51, 59, 62, 69
haemoglobin 3
hairs 21
half-equations 68
halogens 38
hard water 67
hazards, electrical 92
heart disease 29
heat 60, 61, 79, 80, 94, 99
helium nuclei 97
hereditary units 31
hideously important 86
highway code 78
homeostasis 21, 22
hooligan kids 79
household electrics 89
human embryos 29
Huntington's 33
hydrated copper sulfate crystals 60
hydrochloric acid 19, 64-67
hydrogen 42, 62, 68-70
   test for 65
hydrogen peroxide 57
hydrogenated vegetable oil 69
hydroxides 66, 67

I

identical twins 26
immunosuppressive drugs 24
impurities 52
in vitro fertilisation (IVF) 33
independent variable 56, 102
indicator 64, 67
indigestion 64
industrial reactions 51, 59, 62, 69
industry 20
insecticides 69
insoluble bases 67
insoluble salts 67
insulating materials 83
insulin 23, 24
intensive farming 13
inter-molecular forces 43
ion 40, 41, 44
ion content of blood 21, 22

Index

ionisation 97
iron catalyst 59, 62
isomerase 20
isotopes 37, 96

J
joules 79

K
kidneys 22
kilograms 74
kinetic energy 80

L
large intestine 19
laws of motion 75
light dependent resistor (LDR) 87
lightning 84
lime (calcium hydroxide) 64
limiting factors 7, 8
lipase 18-20
lipids 9
litmus paper 69
live wire 92, 93
liver 18, 19, 22
loudspeakers 87
lung cancer 98

M
mains supply 91
male characteristics 30
malleable 44
maltose 18
margarine 69
mass 74, 75, 80, 81
mass balance 74
mass number 36, 47
mean 103
meiosis 28
melting points 40, 43
membrane 27
Mendel, Gregor 31
meniscus 103
merry men 96
metal 44, 65
metal carbonate 67
metal hydroxide 66, 67
metal oxide 66
metallic bonds 44
methane 42
microorganisms 12, 14, 20
mineral deficiency 10
minerals 8, 10
mines 98
mitochondria 2, 17
mitosis 27, 28
mole 50
molecular engineering 45
momentum 81
monoculture 10
moons 74
motion 75, 76
motor, motor effect 87
M_r 47
muscles 17
my little MGB 54

N
naming salts 65
nanomaterials 45
nervous system 33
neutral pH 64
neutral wire 92, 93
neutralisation 60, 64-67
neutrons 36, 96, 99
Newton, Isaac 75
newton meter 74
newtons 74
nitinol 45
nitrate salts 10, 65, 67
nitric acid 65-67

nitrogen 62
noble gases 38
nuclear bombs 99
nuclear decay 96
nuclear fission 99
nuclear fusion 99
nuclear power stations 99
nuclear waste 98
nucleon number 36
nucleus 2, 26, 36, 96

O
offspring 32
OH⁻ ions 64
operating current 93
opposing force 76
orbits 74
ordered variables 104
organic farming 13
organs 2
osmosis 5
outer shell 38
overhead cables 83
oxidation reactions 60
oxides 66, 67
oxygen 3, 6, 7, 17, 42, 57

P
palisade leaf cells 3
pancreas 18, 19, 23
    transplant 24
paper rollers 84
parachute 77
paraffin heater 8
parallel circuits 85, 89, 90
partially permeable membrane 5
paternity testing 26
pea plants 31
penetration, radiation 97
pepsin 16, 18, 19
percentage mass 48
percentage yield 51, 52
periodic table 38, 39, 40
pH 16, 19, 64
phosphates 10
photon 97
photosynthesis 3, 6-8, 14, 17
planets 74
plant growth 10
plastics 69
plugs 92
plum pudding model of atom 96
plutonium 99
poisoning (of catalysts) 59
poisonous ions 67
polythene rod 83
potassium 10
potato cylinders 5
potential difference (P.D.) 85, 88, 89
potential energy 80
power 94
power rating 94
power supply 87, 88
precipitation reaction 67
precision 103, 105
predicted yield 52
prejudice 33
pressure 54, 61
pretty bad news 85
primary consumer 11
probability 30, 34
product 46
protease 18-20
protein synthesis 17
proteins 9, 10, 16-18, 20, 26, 66
proton number 36, 37
protons 36, 37, 96
pyramids of biomass 11, 12
pyramids of number 11

R
radioactive waste 99
radioactivity 96-99
radon gas 98
random error 103
range 103
rate of reaction experiments 56, 57
rates of reaction 54-59, 65
reactant 46
reaction 54-56, 58, 59, 65
reaction forces 76
reactivity 65, 67
reactivity series 70
recessive alleles 31-34
rectum 19
red blood cells 3
reduction with carbon 70
relative atomic mass 47, 48, 50
relative formula mass 47, 48, 50
repelling 83
reproduction 3
reproductive organs 28
repulsion 83
resistance 77, 85, 86, 88, 89, 94
resistors 87
respiration 9, 12, 14, 17, 21
resultant force 75, 76
reversible reaction 52, 60-62
ribosomes 2
rocks 98
rusting 54
Rutherford's Scattering 96

S
saliva 19
salivary glands 18, 19
salt 64-66, 69
satellites 74
scattergram 105, 106
seat belts 81
secondary consumer 11
seeds 9
sensitivity of an instrument 105
semiconductor diode 87
series circuits 85, 88, 90
sex cells 28
sexual reproduction 28
shells 36
shivering 21
shocks, electric 83, 84
sickle cell anaemia 29, 34
silica gel 20
silicon dioxide 43
simple molecules 43, 44
skin 21
small intestine 18, 19
smart materials 45
smoke 84
soap 69
sodium 22, 67
sodium thiosulfate 57
soluble salts 67
sound 79
spark 83, 84
specialised cells 29
speed 72, 73, 76, 80
speed limits 78
speed of reaction 54-59, 65
sperm cells 3, 28
spinal injuries 29
spring balance 74
squeaky pop 65
standard test circuit 85
starch 9, 18
state symbols 46
static charges 83, 84
static electricity 83, 84
stem cells 24, 29
stomach 18, 19
stomata 3

stopping distances 78, 80
streamlined 77
submarines 99
successful collisions 58
sugars 18
sulfate salts 65
sulfur 57
sulfuric acid 65, 66
Sun 98
sunlight 6
surface area 54, 56, 58, 59
sustainable development 59
sweat 21, 22
symbol equation 46
systematic error 103

T
teenagers 38
temperature 16, 54, 56-61
temperature detectors 87
tension in a rope 79
terminal velocity 77
test circuit 85
testing components 85
thermal decomposition 60
thermistor 87
thermoregulatory centre 21
thermostats 87
thinking distance 78
Thomson, J J 96
timebase (oscilloscope) 91
timeless mysteries 36
tissue fluid 5
tissues 2
transferring liquids 52
transition metals 39, 59, 66
tread depth 78
trophic levels 11, 12

U
unbalanced force 75
universal indicator 64
uranium 99
urea 21, 22
urine 22

V
vacuole 2
variable 102, 104-106
variable resistor 85, 87
variation in organisms 27, 28
velocity 72, 73, 75, 81
velocity-time graphs 73
voltage 83-85, 88, 89, 94
voltage-current (V-I) graphs 85, 86
voltmeter 85, 87, 88
volts 85
volume of gas 61
    measuring 55, 56

W
water 5, 6, 17, 42
    test for 60
water bath 8, 57
water content in body 21, 22
weight 74, 76, 79
wild scratty hair 32
wiring a plug 92, 93
work done 79

X
X-chromosome 30

Y
Y-chromosome 30
yield 52, 62

Z
zero resultant force 75